Symbiotic Interactions

Symbiotic Interactions

by

A. E. DOUGLAS

Royal Society University Research Fellow
Department of Biology
University of York

Oxford   New York   Tokyo
OXFORD UNIVERSITY PRESS
1994

Oxford University Press, Walton Street, Oxford OX2 6DP

Oxford   New York   Toronto
Delhi   Bombay   Calcutta   Madras   Karachi
Kuala Lumpur   Singapore   Hong Kong   Tokyo
Nairobi   Dar es Salaam   Cape Town
Melbourne   Auckland   Madrid
and associated companies in
Berlin   Ibadan

Oxford is a trade mark of Oxford University Press

Published in the United States
by Oxford University Press Inc., New York

© Angela E. Douglas, 1994

A catalogue record for this book is available from the British Library

Library of Congress Cataloging in Publication Data

Douglas, A. E. (Angela Elizabeth), 1956–
Symbiotic interactions/by A. E. Douglas.
Includes index.
1. Symbiosis.   I. Title.
QH548.D68   1994   574.5'2482–dc20   93–22183
ISBN 0–19–854286–0 h/b
ISBN 0–19–854294–1 p/b

Typeset by Footnote Graphics, Warminster, Wiltshire
Printed in Great Britain by
Bookcraft (Bath) Ltd
Midsomer Norton, Avon

Preface

Symbiotic interactions arises from an earlier book, *The biology of symbiosis*, written by Sir David Smith and myself. As with *The biology of symbiosis*, this book concerns a coherent group of associations, including corals and other alga–invertebrate symbioses, lichens, cellulose-degrading micro-organisms in the guts of animals, and luminescent bacteria in some marine fish.

Symbiotic interactions is not, however, simply a second edition of *The biology of symbiosis*. This is a new book, with a fundamentally different format and approach. Instead of treating each major group of associations in turn, symbioses are discussed as a series of topics, including nutritional interactions, the formation of symbioses, and their ecological impact. This approach reflects my belief that study across the different symbioses provides a distinctive perspective of the interactions underlying symbioses.

Many biologists have considered symbioses as mutually beneficial associations. As has been recognized in several texts (including *The biology of symbiosis*), this approach is not acceptable, simply because many organisms do not invariably benefit from symbiosis. There has, however, been no general theme to replace the concept of mutual benefit in symbiosis research. Fundamental to *Symbiotic interactions* is that symbiosis is a route by which organisms gain access to novel metabolic capabilities, such as photosynthesis, nitrogen fixation, and cellulose degradation. This theme will be developed through the book. The aim is to provide an integrated approach to symbiosis that will be of value to undergraduate and postgraduate courses, and an aid for researchers in symbiosis.

One of the greatest difficulties in writing this book has been deciding which symbioses to include and which studies to describe. For the sake of clarity and brevity, I have had to exclude many excellent descriptive and experimental investigations. I have, inevitably, omitted the favourite systems of some readers.

Many colleagues have discussed their data and ideas with me, and this has made *Symbiotic interactions* both an easier and a more enjoyable book to write. I am particularly grateful to Dr J. B. Searle and Sir David Smith for their detailed comments on many drafts and their constructive suggestions and encouragement. Professor A. H. Fitter, Dr R. Honegger, Dr M. J. McFall-Ngai, and an anonymous adviser for the Oxford University Press read individual chapters, and Dr R. Law also gave me valuable advice. A number of colleagues kindly provided illustrations: J. Bohatier, C. M.

Cavanaugh, B. J. Finlay, I. B. Heath, D. J. Hill, R. Honegger, R. M. Leech, P. J. McAuley, G. I. McFadden, M. J. McFall-Ngai, R. Parsons, M. Slaytor, T. N. Taylor, and J. R. Turner. I also thank the editors at Oxford University Press for their support.

Finally, I thank Jeremy Searle for his unswerving conviction that I would complete this book.

York   A. E. D.
February 1993

Contents

1

An introduction to symbiosis

1.1 WHAT DOES THE TERM 'SYMBIOSIS' MEAN?

The term 'symbiosis' derives from the Greek for living together, but no definition of symbiosis is universally accepted. The difficulties can be appreciated by considering three associations.

1. *Schistosoma* is a parasitic trematode worm. The adult worm lives in blood vessels of mammals and birds. Humans are host to three species, *S. mansoni*, *S. japonicum*, and *S. haematobium*, and people with schisto-somiasis have a variety of debilitating symptoms, including lassitude and diarrhoea. The World Health Organization has estimated that in 1990 200 million people were infected.

2. Many flowering plants are pollinated by insects. The insect transfers pollen as it visits plants to feed on nectar (as for most butterflies) or nectar and pollen (e.g. honeybees).

3. Aphids are insects which feed on the phloem sap of plants. They require bacteria of the genus *Buchnera*, which are contained within special aphid cells in the body cavity. If the bacteria are eliminated from the aphid (using antibiotics), the aphids fail to grow properly and they die without producing any offspring.

Let us consider which of these three associations are symbioses.

The term 'symbiosis' was first coined by a plant pathologist, Anton de Bary, in the mid-nineteenth century. He defined symbiosis as the living together of different species. De Bary explicitly treated parasitism as a type of symbiosis, but excluded associations of short duration. By this defini-tion, the association between aphids and bacteria and the infection of humans with *Schistosoma* are symbioses, but insect pollination of flowering plants is not.

Some biologists do not accept this interpretation of symbiosis. There are two difficulties. First, it has been argued that interactions with short, even fleeting, contact should be considered as symbioses. By this reasoning, in-sect pollination of flowering plants is a symbiosis. Second, many biologists, especially parasitologists, do not consider parasitic associations, such as the relationship between humans and *Schistosoma*, as symbioses. An even more restrictive view is widespread: that symbioses are associations in which all organisms benefit.

1.2 SYMBIOSIS AND BENEFIT

The view that symbioses should be defined as associations from which all the participating organisms derive benefit has led to a widespread (but rarely explicit) assumption that any association which is not overtly parasitic is mutually beneficial. This raises a question: how can benefit be identified?

The usual method to establish whether an organism benefits from an association is to compare its performance (survival, growth, fecundity, etc.) in the association and in isolation. If the organism performs better in symbiosis, it benefits from the association, but if it performs better in isolation, it is harmed by the association.

It is occasionally stated that benefit is equivalent to enhanced fitness in symbiosis. This is misleading, because benefit and fitness are assessed in different ways. As illustrated in Fig. 1.1, benefit is assessed from the relative performance of one organism in two environments (in association and in isolation); but fitness concerns the performance of two different organisms in the same environment.

Despite the importance attached to benefit in symbiosis, very few de-tailed studies have been conducted to establish whether organisms benefit from associations. Two sets of studies, one on the relationship between hydra and algae and one on mycorrhizal associations between plant roots and fungi, are considered here. They show that many factors, including environmental conditions and developmental age of the organism, may influence whether an organism derives benefit or harm from an association.

(a)   (b)

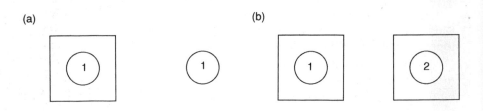

Fig. 1.1 Benefit and fitness in symbiosis. (a) Benefit. Consider an organism (1), which may occur in symbiosis (square) and apart from the symbiosis. If the organ-ism performs better in symbiosis, it benefits from the association. (b) Fitness. Consider two organisms (1, 2), both in symbiosis. Organism 1 is fitter than organ-ism 2 if it performs better than organism 2 under the same environmental con-ditions of symbiosis.

1.2.1 **Benefit in the green hydra symbiosis**

Hydra are freshwater Cnidaria, living in ponds and slowly moving rivers, where they feed on small animals, such as water fleas and rotifers, in the water column. Some hydra are green because they contain algae of the genus *Chlorella* in their cells (Fig. 1.2). The algal cells can photosynthesize, and they release substantial amounts of photosynthetically fixed carbon to the animal cells in the form of a sugar, maltose.

Green hydra are easy to maintain in the laboratory, and they can be 'bleached' of their algae by incubation at very high light intensity. The hydra containing algal cells (symbiotic hydra) and those deprived of their algae (aposymbiotic hydra) reproduce rapidly by asexual budding. Large numbers of genetically identical symbiotic and aposymbiotic hydra can thus be generated for experiments.

To investigate whether green hydra benefit from their symbiosis with *Chlorella*, the performance (survival and growth) of symbiotic and aposym-

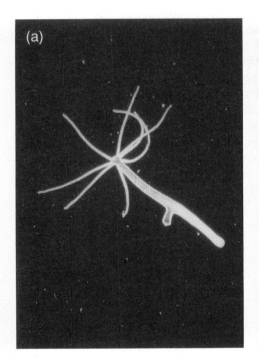 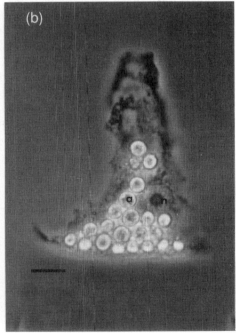

Fig. 1.2 The green hydra–*Chlorella* symbiosis. (a) The hydra. (b) An isolated digestive cell of the hydra containing approximately 20 cells of the alga *Chlorella* (a). The hydra cell nucleus (n) is also a conspicuous in these cells. Scale: 8 μm. (Micrographs provided by P.J. McAuley).

biotic hydra are compared under different environmental conditions (Table 1.1).

When the hydra are maintained in the light without food, the aposymbiotic hydra die within a few weeks. Under these conditions, the symbiotic hydra decline in size but they persist and, even after 2–3 months without food, they respond to food presented to their tentacles, digest the food, and initiate growth. It appears that green hydra benefit from the symbiosis when food is scarce, provided they are illuminated. The sugars derived from algal photosynthesis presumably enable green hydra to tolerate starvation.

A very different picture of the green hydra symbiosis emerges when animals are fed. Well-fed aposymbiotic and symbiotic hydra grow at closely similar rates in the light. The algae continue to release maltose, but this source of nutrients is of no significance to the animal. Fed cultures of the hydra can also be maintained indefinitely in the dark. The algal population in symbiotic hydra declines by about 60 per cent, but the animals remain pale-green. In darkness, however, the symbiotic hydra grow significantly more slowly than aposymbiotic hydra, probably because nutrients derived from the food are diverted to maintain *Chlorella* which, in the absence of photosynthesis, contribute nothing to hydra.

In summary, environmental conditions are a major factor determining whether green hydra benefit from possessing *Chlorella*. The hydra do benefit (in terms of enhanced survival) when food is scarce, but *Chlorella* are of no discernible value to fed animals in the light; and they are detrimental, reducing the growth of fed hydra, in the dark.

Table 1.1 Performance of green hydra under different conditions of illumination and feeding

Hydra	Starved in light (Survival)	Fed in light (population increase, number of individuals produced per individual per day (mean ± s.e.))	Fed in dark
Symbiotic	> 10 weeks	0.10 ± 0.004	0.07 ± 0.003
Aposymbiotic	< 2 weeks	0.09 ± 0.002	0.09 ± 0.003
	The hydra benefit	The hydra derive neither benefit nor harm	The hydra are harmed

(Data from Douglas and Smith (1983)).

1.2.2 **Mycorrhizal fungi and plant performance**

The roots of many plants bear fungi, known as *mycorrhizal fungi*. The fungal partner obtains its carbon requirements from the plant, and it enhances the plant's capacity to take up mineral nutrients, especially phosphate, from the soil. A plant may benefit from the association if the advantage of improved mineral nutrition outweighs the cost of providing photosynthetic carbon to the fungus.

It has repeatedly been demonstrated that mycorrhizal fungi improve the performance of plants grown in the laboratory or glasshouse (Harley and Smith 1983). The advantage to the plant is particularly evident on nutrient-poor substrata. For example, in the study of Koucheki and Read (1976) on the grass *Festuca ovina*, the mycorrhizal fungus significantly enhanced the growth rate of the plant on the lowest two of the four nutrient treatments used (Table 1.2). At the higher concentrations of applied nutrients, the performance of plants with and without the mycorrhizal fungi was similar.

In a minority of laboratory studies, the mycorrhizal fungi reduce plant performance. In particular, seedling growth may be retarded. Photosynthetic carbon is often limiting in young seedlings, and when these seedlings are infected with a mycorrhizal fungus the photosynthate is diverted from plant growth to the fungus.

Most laboratory experiments are conducted on plants grown singly in pots. The results cannot be extrapolated to natural conditions, where the relationship between plants and mycorrhizal fungi is considerably more complex. Established vegetation supports a network of mycorrhizal fungi.

Table 1.2 The effect of mycorrhizal fungi on the growth of the grass *Festuca ovina*
Four-week old plants were inoculated with the mycorrhizal fungus, and both the infected plants and fungus-free control plants were grown for 4 months in a peat-sand substrate with differing levels of nutrients applied every third day: no nutrients (0); 7 ppm N and 4 ppm P (1); 14 ppm N and 8 ppm P (2), 28 ppm N and 16 ppm P (4)).

Condition of plants	Mean relative growth rate of plants (mg g^{-1} d^{-1})			
	4	2	1	0
Mycorrhizal	2.94	2.73	2.27	1.24
Non-mycorrhizal	2.87	2.56	1.16	0.76
	$p > 0.05$	$p > 0.05$	$p < 0.01$	$p < 0.01$

(Data from Koucheki and Read (1976)).

For one type of mycorrhizal fungus, the vesicular-arbuscular mycorrhizal fungi, in particular, hyphae from a single fungal mycelium may associate with the root systems of many different plants, while each plant may interact with several different species of fungi (Fig. 1.3). In this system, the fungus is believed to derive most photosynthate from plants with the highest photosynthetic capacity, and the plant with greatest phosphorus demand gains most phosphate from the fungal mycelium.

The limitations of the laboratory experiments can be illustrated by the response of seedlings to mycorrhizal fungi. In established vegetation, the developing root systems of seedlings can 'tap into' a pre-existing mycorrhizal association with fungi whose carbon needs are already met by other plants. The seedling can therefore gain mineral nutrients for very little cost (in terms of photosynthate to the fungus). Mycorrhizal fungi may be particularly important for seedling establishment in field conditions. As already mentioned, the opposite conclusion is reached from laboratory experiments, in which a seedling has to meet the entire carbon needs of its fungal partner.

In summary, both environmental conditions (specifically, the concentration of phosphorus in the soil) and developmental age of the plant (seedlings or mature plants) can influence whether a plant benefits from the association with mycorrhizal fungi. Detailed aspects of the plant–fungal relationship can also be important. Seedlings supporting a fungal mycelium are harmed by the association, but those that form an association with an established plant–fungal network (as in Fig. 1.3) benefit from the association.

1.2.3 Benefit and the meaning of the term *symbiosis*

Both the hydra–*Chlorella* association and the relationship between plants and mycorrhizal fungi are commonly regarded as symbioses. Hydra and

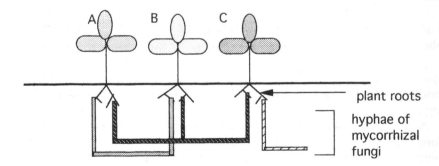

Fig. 1.3 The relationship between plant roots and mycorrhizal fungi in natural vegetation. Plants A, B, and C are linked by hyphae from three distinct fungi.

plants do not, however, always benefit from the associations. In other words, benefit is not an invariant property of an association, but is dependent on the interaction between the association and its environment.

Returning to the various meanings of symbiosis considered in section 1.1, it is clearly inappropriate to include benefit as a defining characteristic of symbioses. I favour the original definition of de Bary: symbioses are associations between different species that persist for long periods (relative to the generation time of the interacting organisms).

1.3 SYMBIOSIS AS A SOURCE OF NOVEL METABOLIC CAPABILITIES

This book concerns a coherent group of symbioses, including lichens, corals and other alga–invertebrate symbioses mycorrhizal associations between plant roots and fungi, and cellulose-degrading microorganisms in the guts of herbivorous animals. The associations are listed in Table 1.3. They are traditionally considered as mutually beneficial symbioses (for example, Buchner 1965) but, as considered above, the organisms do not always benefit from the association.

The theme of this book is that the common denominator of the associations is not mutual benefit but a novel metabolic capability, acquired by one organism from its partners.

Let us consider the symbioses in Table 1.3. Every symbiosis includes a eukaryotic organism. Eukaryotes are defined as organisms with a double-membrane bound nucleus, as distinct from the eubacteria and archaebacteria in which the genome is not separated from the cytoplasm by a membrane (see Table 1.4). The eukaryotes are the only group with major radiations of multicellular organisms (the animals, the terrestrial plants, and the fungi). However, despite their remarkable morphological diversity and complexity, the eukaryotes have very restricted metabolic capabilities. The lineage giving rise to the eukaryotes could not respire aerobically, photosynthesize, or fix nitrogen, and some eukaryotic groups have subsequently lost a variety of capabilities. For instance, the animals can synthesize neither 10 of the 20 amino acids that make up proteins (the essential amino acids) not many coenzymes required for normal functioning of enzymes central to metabolism (these include the B vitamins).

A number of eukaryotes have overcome these metabolic limitations by forming symbioses with other organisms that possess the appropriate biochemical capabilities. The ability of almost all eukaryotes to respire aerobically derives from a symbiosis with Proteobacteria which evolved into mitochondria. Similarly, photosynthesis by algae and plants is mediated by plastids, which have evolved from cyanobacteria. (Both Proteo-

Table 1.3 Symbiosis as a source of novel metabolic capabilities

Capability acquired by symbiosis	Donor of capability	Recipient of capability
Photosynthesis	Algae and cyanobacteria	Various protists, invertebrates and lichenized fungi
	Terrestrial plants	Mycorrhizal fungi
Chemosynthesis	Bacteria	Various invertebrates
Nitrogen fixation	Rhizobia, *Frankia*, cyanobacteria	Plants
	Various bacteria	Some termites
Nutrients, e.g. essential amino acids, vitamins	Various bacteria	Many animals especially insects on nutrient-poor diets
Methanogenesis	Methanogenic bacteria	Anaerobic protists
Cellulose degradation	Bacteria, e.g. *Ruminococcus*	Vertebrates, especially herbivorous mammals
	Hypermastigote protists	Lower termites
Luminescence	*Vibrio* and *Photobacterium*	Marine cephalopods and teleost fish

Table 1.4 The major divisions of living organisms

Molecular studies (summarized in Woese (1987) and Woese *et al.* (1990)) have identified three major domains of living organisms: the Eucarya, Eubacteria, and Archaea (also known as archaebacteria). The Eubacteria, Archaea, and protists (unicellular eukaryotes) are known as micoorganisms.

Domain	Organisms
Eucarya	The eukaryotes, comprising protists[1], fungi[1], animals, and plants
Eubacteria	Most bacteria, including the Proteobacteria (comprising α, β, γ and δ groups)
Archaea	The archaebacteria, including methanogenic bacteria and some extreme halophiles and thermophiles.

[1] The protists are groups of eukaryotes that are predominantly unicellular. Algae are protists with plastids.
[2] The 'true' fungi, i.e. Zygomycotina, Ascomycotina and Basidiomyotina (and asexual Deuteromycotina).

bacteria and cyanobacteria are eubacteria; see Table 1.4.) Other novel metabolic capabilities acquired by eukaryotes include: nitrogen fixation, gained by many leguminous plants (e.g. pea, clover) through an association with nitrogen-fixing bacteria called rhizobia; essential amino acids, acquired by aphids from the symbiosis with bacteria (described in section 1.1); and cellulose degradation gained by most herbivorous mammals from cellulolytic microorganisms in their guts.

1.4 HOSTS, SYMBIONTS, AND THE LOCATION OF SYMBIONTS

Some basic terminology is essential to understanding symbiosis, and it is introduced in this section.

For most of the associations described in this book, one organism is considerably larger than its partner(s). It is known as the *host*, and the smaller organisms are called *symbionts*. Usually, the host is the recipient of the novel metabolic capability. (The associations between plant roots and mycorrhizal fungi are unusual in that both partners are eukaryotes of considerable size [the fungus inhabits both roots and the surrounding soil]. The terms 'host' and 'symbiont' are not appropriate to these symbioses.)

Most symbionts are located within the body of their host. Many are extracellular, i.e. external to cells of the host. These include bacteria and protists in outpocketings of animal guts, and the photosynthetic symbionts (algae or cyanobacteria) of lichens (Fig. 1.4(a)). Intracellular symbionts are internal to the cell membrane of host cells, usually in the cytoplasm. The intracellular symbionts are invariably acquired by endocytosis, and most remain separated from the host cell contents within an organelle, called the *symbiosome* (Fig. 1.4(b)). The photosynthetic algae in corals and the bacteria in aphids are examples of intracellular symbionts.

1.5 SCOPE OF THE BOOK

The single most important point raised in this introductory chapter is that symbiosis is a major route by which eukaryotes have gained access to complex metabolic capabilities. An introduction to the relevant symbioses is provided in Chapter 2.

It will become apparent that symbiosis has had a dramatic impact on the morphology of some organisms, and that certain symbiotic structures are of remarkable anatomical complexity. The structural aspects of symbioses are reviewed in Chapter 3. Among the examples considered are the thalli of certain lichens, which are among the most complex structures to have

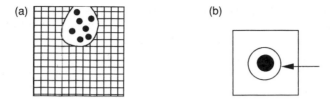

Fig. 1.4 Extracellular and intracellular symbionts. Most symbionts (●) are located in the body of their host. (a) Some multicellular hosts bear extracellular symbionts in a cavity (as illustrated) or, less commonly, between closely apposed cells. Unicellular hosts may have extracellular symbionts between the cell wall and cell membrane. (b) An intracellular symbiont is within a host cell (indicated by the square). Intracellular symbionts are invariably acquired by endocytosis, and most remain separated from the host cell contents by a derivative of the endocytotic membrane (arrow), which is known as the *symbiosome membrane*.

evolved in fungi; and certain fish which, linked to their possession of luminous bacteria, have lenses and mirrors that control the emission of light.

Some of the mechanisms underlying symbioses as a source of novel metabolic capabilities are considered in Chapters 4–6:

1. How does the eukaryotic recipient gain access to the metabolic capability of its partner (Chapter 4)? As an example, let us consider the nitrogen-fixing rhizobia in root nodules of leguminous plants. Virtually all the nitrogen fixed by the rhizobia ia released to the plant. Somehow, the rhizobia are induced to contribute to the nitrogen requirements of the plant, instead of channeling all the products of nitrogen fixation into their own proliferation.

2. How is the appropriate donor of the metabolic capability selected (Chapter 5)? Continuing with the rhizobia–legume example, it is crucial to the legume that it selects an appropriate nitrogen-fixing bacterium which provides fixed nitrogen, and not one that offers little fixed nitrogen, possibly while utilizing plant nutrients.

3. How does a stable association persist, so that no participating organism overgrows its partners and the various organisms do not separate (Chapter 6)? Some associations are remarkably long-lived; for instance, certain lichen thalli may persist for hundreds of years. The maintenance of a stable association involving multicellular eukaryotes and microorganisms is particularly interesting because microorganisms, in general, can proliferate much more rapidly than animals or plants. As an example, the bacteria (*Buchnera*) in aphids are related to *Escherichia coli*, which is capable of a doubling time of 20 minutes, but *Buchnera* increases at similar rates to the insect, with a doubling time of 3–4 days.

In addition to these processes underlying the establishment and maintenance of symbioses, there are immense biological consequences of symbiosis. Symbiotic systems dominate certain habitats (Chapter 7). In particular, much of the terrestrial vegetation bears symbiotic fungi or bacteria. These associations can contribute to the capacity of plants to colonize barren soil, and they influence both competitive interactions between plant species and the dynamics of plant succession.

Without doubt, the most significant associations in the evolutionary history of eukaryotes have been those with the symbiotic bacteria that gave rise to mitochondria and plastids. Which group of bacteria did the mitochondria and plastids evolve from, and what was the host cell in which these organelles arose? How do mitochondria and plastids differ from symbionts, and how are symbionts transformed into organelles? These issues are discussed in Chapter 8.

2

Symbiosis as a source of novel metabolic capabilities

As considered in Chapter 1, eukaryotes have obtained access to a number of metabolic capabilities through symbiosis. The principal associations are listed in Table 1.3. In this chapter, they are described in terms of the metabolic capability acquired: photosynthesis and chemosynthesis, nitrogen fixation, the synthesis of vitamins and essential amino acids, methanogenesis, cellulose degradation, and luminescence.

The information in this chapter can be used both as an introduction to the symbioses and as a source of reference for later chapters, where symbioses are considered in various different contexts.

2.1 CARBON DIOXIDE FIXATION: PHOTOSYNTHESIS AND CHEMOSYNTHESIS

Sugars and other organic compounds can be synthesized from carbon dioxide, the energy and reductant being provided either by the trapping of light on to chlorophyll pigments (photosynthesis) or by the oxidation of reduced inorganic compounds, such as sulphide (chemosynthesis).

Eukaryotes have gained the capacity for photosynthesis and chemosynthesis only through symbiosis. The plastids in algae and plants are derived from photosynthetic symbionts (see Chapter 8). A number of terrestrial fungi and aquatic invertebrates and protists have photosynthetic algae or cyanobacteria, and certain invertebrate animals possess chemosynthetic bacterial symbionts. These symbioses are however, unknown in vertebrates or arthropods.

2.1.1 **Photosynthetic associations in aquatic protists and invertebrates**

Photosynthetic symbionts occur in protists (e.g. Ciliates, Foraminifera, and Radiolaria) and a few phyla of animals (Table 2.1). The most abundant animal hosts are the Cnidaria, and virtually all species in tropical shallow waters are symbiotic. They include stinging corals (Millepora), rhizostome jellyfish (e.g. *Cassiopea*), and a great many stony corals (Scler-

Table 2.1 Associations of protists and invertebrates with photosynthetic algae and cyanobacteria

(a) **Marine systems** Hosts	Symbionts
Protists	
Foraminifera	Various algae, e.g. chlorophytes, diatoms
Radiolaria	Dinoflagellates
Acantharia	Not identified
Invertebrates	
Porifera (sponges)	Cyanobacteria or algae
Cnidaria	Dinoflagellates, usually *Symbiodinium*
Mollusca (some gastropods and bivalves including tridacnid clams	Dinoflagellates, usually *Symbiodinium*
Turbellaria (flatworms)	Various algae, e.g. diatoms, *Tetraselmis*
Ascidiacea (sea squirts)	Cyanobacteria, often *Prochloron*

(b) **Freshwater systems**
Chlorella is the usual symbiont in freshwater protists (many ciliates and some amoebae) and animals, especially demosponges, hydras (Cnidaria), neorhabdocoel flatworms, and the bivalve molluscs *Anodonta* and *Unio*.

actinia), octocorals, zoanthids, and sea anemones (Fig. 2.1(a)). The stony corals and stinging corals are of immense ecological importance as the main framework builders of coral reefs; and the capacity of these animals to generate calcareous skeletons fast enough to withstand wave action in shallow waters can be linked directly to photosynthesis by their symbiotic algae.

The symbiont in most marine Cnidaria is a gymnodinoid dinoflagellate alga, *Symbiodinium* (Fig. 2.1(b)). *Symbiodinium* is usually intracellular, surrounded by multiple symbiosome membranes of animal origin. It also occurs in other animals, notably the tridacnid clams.

A variety of other photosynthetic symbionts are known in marine hosts. For example, marine sponges may have diatoms, cryptophytes, dino-flagellates, and cyanobacteria; and dinoflagellates, chlorophytes, diatoms, and rhodophytes have been identified in a major group of protists, the Foraminifera.

The symbionts in most freshwater protists and invertebrates are algae of

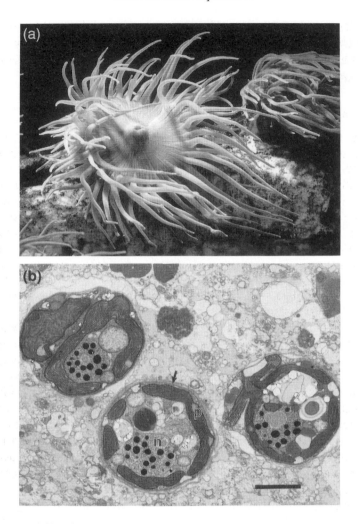

Fig. 2.1 Algal symbiosis in Cnidaria: the sea anemone *Anemonia viridis* and the dinoflagellate alga *Symbiodinium*. (a) *Anemonia viridis* (photograph: J.R. Turner). (b) Cells of *Symbiodinium* sp. in *A. viridis*. The algal cells are coccoid, with permanently condensed chromosomes in the nucleus (n) and a peripheral plastid (p). Each *Symbiodinium* cell is separated from the animal by single or multiple symbiosome membranes (arrow). Scale: 4 μm. (Micrograph: A.E. Douglas)

the *Chlorella vulgaris* group. The algae are intracellular in the protists and hydras (freshwater Cnidaria) (see Fig. 1.2), but it is uncertain whether they are intracellular or extracellular in the freshwater sponges, flatworms, and bivalve molluscs.

2.1.2 The lichen symbioses

Lichens are associations of fungi with algae or cyanobacteria. Most are terrestrial. They dominate large areas of tundra and many high-altitude habitats, and they are also abundant in temperate and tropical forests (Kappen 1988). Lichens can grow on rock, bark, undisturbed soil and certain man-made structures (e.g. concrete). Their principal requirements are a stable substratum (because they grow very slowly) and illumination. Lichens can tolerate desiccation and extremes of temperature, but their symbionts are photosynthetically active only when the lichen is moist.

In all lichens, the photosynthetic algae or cyanobacteria are enveloped by the fungus and are extracellular. Lichens may vary from crust-like or globular structures to complex thalli, as in the foliose and fruticose lichens (Fig. 2.2).

In almost all of the 13 500 species of lichens, the fungal host belongs to the Ascomycotina, and lichens represent nearly half of all ascomycotine species (Table 2.2(a)). Other lichenized fungi are a few Basidiomycotina (all Hymenomycetales) and Deuteromycotina (fungi with no sexual stages: some are allied with the lichenized Ascomycotina). The symbionts in lichens include 30–40 genera of algae, predominantly of the Chlorophyta, and at least 12 genera of cyanobacteria (Table 2.2(b)). The most common photosynthetic symbiont is the chlorophyte *Trebouxia*, recorded in more than half of all lichens. Between 20 and 40 per cent of lichens have other chlorophytes, notably *Trentepohlia*, and approximately 10 per cent have cyanobacteria, usually either *Nostoc* or *Scytonema*.

Lichens have evolved many times, and some may be very ancient. Biogeographical studies of lichens in the southern hemisphere suggest that certain genera and even species of the Peltigerales and Pertusiales may have been distinct by the Triassic (Hawksworth and Hill 1984). Other lichens, especially in the Basidiomycotina, undoubtedly have more recent origins.

2.1.3 Mycorrhizas

Mycorrhizas are persistent associations between fungi and roots (or other underground parts) of plants. Most vascular plants can form mycorrhizas (Harley and Smith 1983), and some bryophytes also bear mycorrhizal fungi (Pocock and Duckett 1985).

Fig. 2.2 Lichen thalli. (a) *Rhizocarpon geographicum*, a crustose lichen; (b) *Parmelia caperata*, a foliose lichen; (c) *Cladonia polydactyla*, comprising basal squamules and vertical fruticose structures called podetia; (d) *Usnea* sp., a fruticose lichen on a tree trunk (note foliose lichens also present). (Photographs: D.J. Hill)

Table 2.2 Lichenized fungi and their photosynthetic symbionts

(a) **The fungal hosts** Subdivision of Fungi	Number of lichenized species (%)[1]
Ascomycotina	13 250 (46)
Basidiomycotina	50 (0.3)
Deuteromycotina	200 (1)

[1] Lichenized fungi as percentage of total number of species in each subdivision shown in parentheses.

(b) **The most common symbionts in lichens**

Chlorophyte algae	Cyanobacteria
Order Chlorococcales	Order Hormogonales
Trebouxia	*Nostoc*
Myrmecia	*Scytonema*
Coccomyxa	

Order Trentepohliales
Trentepohlia

Between 30 and 40 genera of algae and cyanobacteria have been reported in lichens, but a comprehensive list of lichen symbionts cannot be constructed, because the symbionts have rarely been studied to the species level and the taxonomy of many of the relevant groups is confused (Hawksworth 1988).

There are four major types of mycorrhizas in vascular plants: vesicular-arbuscular mycorrhizas, ectomycorrhizas, ericoid mycorrhizas, and orchid mycorrhizas (Table 2.3). Nutrient flux in mycorrhizas is usually bidirectional: photosynthetically derived carbon compounds from plant to fungus, and mineral nutrients from fungus to plant. The orchid mycorrhizas are most unusual in that the plant derives carbon from the fungus, and not vice versa as in other mycorrhizas.

The vesicular-arbuscular mycorrhizas (usually known as VA-mycorrhizas) are very abundant, especially in tropical forests, savannah, and temperate grasslands (Fig. 2.3). The ectomycorrhizas become increasingly abundant in temperate and boreal forests especially those dominated by beech (Fagaceae) and pine (Pinaceae). At still higher latitudes or altitudes, ericaceous plants with ericoid mycorrhizal fungi occupy large areas of acidic heathland soils, especially above the tree-line.

At least 75 per cent of all vascular plant species bear VA-mycorrhizas. These include many trees, shrubs, and herbs in most families of angiosperms, and many gymnosperms and pteridophytes (Trappe 1987). All the

Table 2.3 The plants and fungi in mycorrhizas

	Type of mycorrhiza			
	Vesicular-arbuscular	Ecto-	Ericoid	Orchid
Plant	Many groups of pteridophytes, gymnosperms, and angiosperms	Angiosperms and family Pinaceae in gymnosperms	Ericales	Orchidaceae
Fungus	Endogonaceae (Zygomycotina)	Basidiomycotina + Ascomycotina	Ascomycotina	Basidiomycotina

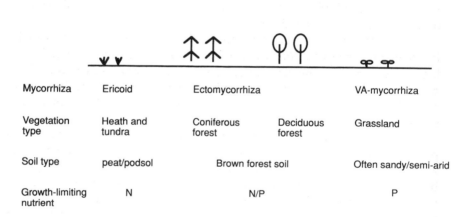

Fig. 2.3 The mycorrhizas associated with different types of vegetation. The latitudinal zones are not sharply defined, and two types of mycorrhiza may occur in one habitat (e.g. in temperate deciduous woodland, many mature trees are ectomycorrhizal while the understorey plants have VA-mycorrhizas). There are also several exceptions to the general scheme in the diagram, such as the abundant VA-mycorrhizas in some alpine and arctic habitats, and the ectomycorrhizal Dipterocarpaceae which dominate some forests in south-east Asia. (Redrawn after Read (1984))

fungi in VA-mycorrhizas are members of the family Endogonaceae in the Zygomycotina. The principal site of nutrient exchange in these mycorrhizas is the arbuscule, a densely branching fungal projection, produced in the inner cortex of the root (Fig. 2.4(a), (b). The fungal hyphae in the root also produce coils and lipid-rich vesicles, but the functions of these structures are uncertain. The fungal hyphae in the root are linked directly to a network of hyphae which extend a considerable distance into the soil. Many VA-mycorrhizal fungi produce large chlamydospores in the soil.

Plant roots bearing ectomycorrhizal fungi are distinctive. They usually have very few, or no, root hairs, and they tend to be short and thick. In most ectomycorrhizas, the root apex is enclosed in a fungal sheath up to 50 μm thick (Fig. 2.4(c)), and fungal hyphae extend from the sheath into the soil. Within the root, the fungus forms a complex branching structure, known as the Hartig net, which mediates nutrient transfer between fungus and plant. Ectomycorrhizas are known in species from 45 families of gymnosperms and dicotyledonous angiosperms (but from no monocotyledons). Most ectomycorrhizal plants are woody perennials in highly seasonal climates; e.g. forest trees at high latitudes (see Fig. 2.3). Thousands of ectomycorrhizal fungal species have been reported. They include an estimated 5000 basidiomycotine species (in the Hymenomycetales and Gasteromycetales), hundreds of ascomycotine species (in the Pezizales, Helotiales, and Elaphomycetales), and many reproductively sterile mycelia of uncertain taxonomic affinity (Harley and Smith (1983).

The ericoid mycorrhizas occur only in the Ericales, and they usually infect plants, such as *Calluna* and *Rhododendron*, living in acidic heathlands (Fig. 2.3). The fungus is ascomycotine, and *Hymenoscyphus* (=*Pezizella*) *ericae* has been isolated into culture. The fungus infects thin lateral roots (known as hair roots), around which it forms a dense mycelium (Read 1983).

The orchid mycorrhizas are restricted to the family Orchidaceae, comprising some 20 000 species. All orchids bear mycorrhizal fungi (Fig. 2.4(d)), at least during the early stages of development. The fungi are basidiomycotine, and several genera have been identified, including *Ceratobasidium* and *Armillaria*. The orchid mycorrhizas are unusual in that the plant is infected at, or soon after, germination, and before the root system develops. The germinating orchid seedling derives organic nutrients from its fungal partner. The orchids that remain partly or completely achlorophyllous (nonphotosynthetic) obtain carbon from the mycorrhizal fungi throughout their lives.

The various mycorrhizal associations probably evolved at the time of origin of their plant partners: the ectomycorrhizas with the gymnosperms in the late Palaeozoic, the orchid mycorrhizas with the Orchidaceae in the Cretaceous, and the ericoid mycorrhizas with the Ericaceae in the early

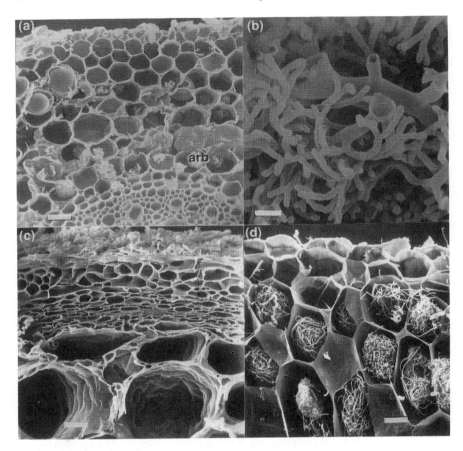

Fig. 2.4 The structural organization of mycorrhizas. (a) Transverse section through root of *Zea mays*, infected with VA-mycorrhizal fungus *Glomus mossae*. Cells of the inner cortex of the root bear fungal arbuscules (arb). Scale: 30 μm. (b) Arbuscules of *Glomus mossae* in a root of *Zea mays*. Scale: 2 μm. (c) Transverse section through root of *Pinus* bearing an ectomycorrhizal fungal sheath (f). Scale: 10 μm. (d) Transverse section through root of the orchid *Neottia nidus-avis*, with an extensive infection of a mycorrhizal fungus in the cortex. The fungal projections are called pelotons (p). Scale: 5 μm. (Scanning electron micrographs: R. Honegger)

Tertiary. The evolutionary origins of the VA-mycorrhizas are uncertain. The earliest unambiguous record of a VA-mycorrhiza is in the Triassic cycad *Antarticycas* (Fig. 2.5(a)) (Stubblefield *et al.* 1987), but fungi (which may be mycorrhizal) have been reported in many Palaeozoic plants (Fig. 2.5(b)) (Stubblefield and Taylor 1988). Pirozynski and Malloch (1975) have argued persuasively that the earliest land plants in the late Silurian

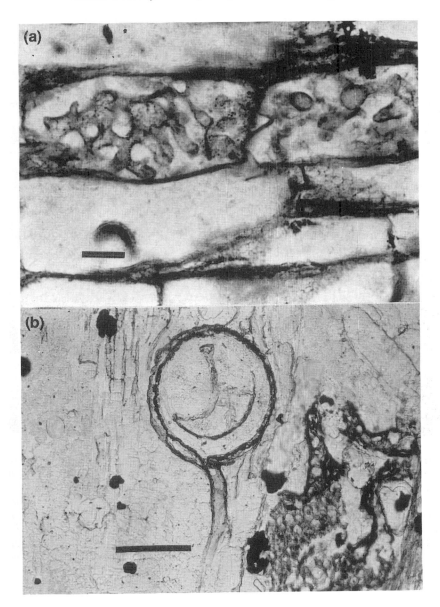

Fig. 2.5 Fossil mycorrhizal fungi. (a) Root of Triassic cycad *Antarticycas* containing highly branched arbuscules of a VA-mycorrhizal fungus. Scale: 15 μm. (Reproduced from Stubblefield *et al*. 1987) (b) Stalked chlamydospore of fungus morphologically similar to VA-mycorrhizal fungus *Glomus*, in cortical tissue of the underground axis of the Carboniferous vascular plant. Scale: 100 μm. (Reproduced from Stubblefield and Taylor (1988))

bore mycorrhizal fungi (perhaps akin to modern VA-mycorrhizal fungi) which enhanced their acquisition of phosphate from the soil. These first land plants would otherwise have faced acute problems in acquiring minerals because they lacked absorptive roots and the soils they utilized were of lower quality than modern soils.

2.1.4 Chemosynthesis in invertebrate animals

Chemosynthetic bacteria require both reduced inorganic compounds and molecular oxygen to generate the energy and reducing power for carbon dioxide fixation, and they are therefore restricted to regions of contact between oxic and anoxic environments. Their most widespread habitat is in marine sediments, at the zone where the oxygen-rich sea water percolating down into the sediment meets the lower anoxic sediment water. Other zones of mixing between reducing and oxic water include the immediate environs of sewage outfalls and pulp-mill effluents, natural gas and oil seeps (e.g. in the Gulf of Mexico), and the deep-sea hydrothermal vents. Most chemosynthetic symbioses are very difficult to study. The hydrothermal vent systems are at a depth of 1500–3500 m, and most other symbioses are sublittoral, often at depths greater than 50 m. Exceptionally, a few bivalves, such as *Lucinoma borealis* on the Atlantic coast of Europe, extend to the lower reaches of the intertidal zone.

Much of the research on chemosynthetic symbioses has been conducted on the hydrothermal vent systems, where these associations were first recognized in the early 1980s. Hydrothermal vents occur in regions of seafloor spreading, i.e. where the continental plates are moving apart. At these zones, sea water seeps into the Earth's crust, where it is heated, absorbs minerals, and its constituent sulphate is reduced to sulphide. The hydrothermal vents are fissures in the sea floor from which the modified 'vent' water is ejected. Associated with the vents is a remarkable community of invertebrate animals, including pogonophorans, e.g. *Riftia pachyptila*, massive bivalves (both clams and mussels), and a variety of polychaete worms and crabs. Some of the animals, including all the pogonophorans and many molluscs, have chemosynthetic symbionts, fuelled by the high sulphide concentrations in the vent waters (Cavanaugh *et al.* 1981).

The symbioses are best studied in the bivalve molluscs and Pogonophora (Table 2.4). In the bivalve molluscs, the chemosynthetic symbioses have evolved independently at least four times: once in the deposit-feeding Protobranchia (*Solemya*) and three times in the filter-feeding Eulamellibranchia (in the mussels and various clams). The bacterial symbionts are located in the gills (Fig. 2.6(a)), which are larger and more 'fleshy' than in nonsymbiotic bivalves. The Pogonophora are related to the annelid worms

Table 2.4 Animals with chemosynthetic symbionts

Host taxa	Habitat
Pogonophora	
Vestimentifera	Hydrothermal vents
(e.g. *Riftia*)	
Perviata	Sediments of
(e.g. *Siboglinum*)	continental shelf
Bivalve molluscs	
Solemyidae	Sewage outfalls,
(e.g. *Solemya*)	mudflats
Mytilidae	Hydrothermal vents,
(e.g. *Bathymodiolus*)	hydrocarbon seeps
Lucinidae	Subtidal sediments
(e.g. *Lucinoma*)	
Vesicomycidae	Hydrothermal vents
(e.g. *Calyptogena*)	
Thyasiridae	Subtidal sediments
(e.g. *Thyasira*)	
Nematodes	
Stilborematinae	Subtidal sediments

Chemosynthetic bacteria have also been reported in the gastropod mollusc *Alviniconcha hessleri* at hydrothermal vents and in the tubificid annelid *Phallodrilus* and a ciliate protist *Kentrophoros*.

and are divisible into two groups: the Vestimentifera, found at hydrothermal vents, and the Perviata, abundant in sediments on the continental shelf and to depths of 2000 m. All Pogonophora have chemosynthetic symbionts located in gelatinous tissue, the trophosome, which fills much of the animal's trunk (Fig. 2.6(b)). Chemosynthetic bacteria have also been documented in a few interstitial animals, including annelids, nematodes and possibly turbellarians, and one protist, the ciliate *Kentrophoros* (Fisher 1990).

As yet, none of the chemosynthetic symbionts have been brought into axenic culture. Most are sulphur-oxidizers, but perviate pognophorans and bivalves of the genus *Bathiomodiolus* probably have methylotrophic forms (Cavanaugh 1992).

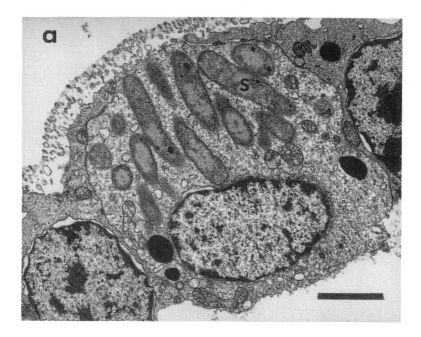

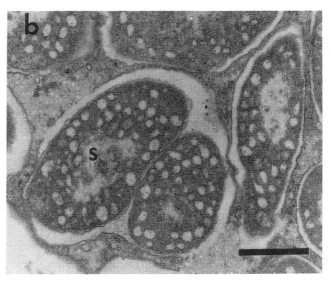

Fig. 2.6 Chemosynthetic symbionts in invertebrate animals. (a) Rod-shaped symbiotic bacteria (s) in cells of the gills of the bivalve mollusc *Solemia velum*. Scale
3 μm. (b) bacteria (s) in cells of the trophosome of the vestimentiferan pogo-
nophoran *Riftia pachyptila*. Scale: 1 μm. (Transmission electron micrographs of
C.M. Cavanaugh, reproduced from Cavanaugh (1983) *Nature*, **302**, 58–61.

2.2 NITROGEN-FIXING SYMBIOSES

Nitrogen fixation is the reduction of dinitrogen (N_2) to ammonia. The reaction can be expressed:

$$N_2 + 3H_2 \rightarrow 2NH_3$$

The reaction is relatively uniform at the enzymological level, catalysed by the nitrogenase complex, which comprises two enzymes: a dinitrogen reductase and a dinitrogenase. The genes necessary for the synthesis of the nitrogenase complex are called *nif* genes.

Nitrogen fixation is believed to be a very ancient capability. It has probably evolved only once, and it is widely distributed among bacteria but absent from eukaryotes. Eukaryotes have gained nitrogen fixation exclusively by symbiosis.

A major factor limiting the incidence of nitrogen fixation is that this process is energetically costly. Nitrogenase is a large and slow enzyme; it is therefore expensive to synthesize, and it has a high demand for ATP (at least 25 mols ATP per mol N_2 fixed). As a result, organisms utilize N_2 only if it is the sole nitrogen source available. In the presence of a combined nitrogen source (e.g. ammonia), the *nif* genes are repressed and nitrogenase is undetectable in the cell. Compounding the energetic cost of nitrogen fixation, nitrogenase is inactivated by molecular oxygen. In other words, nitrogen-fixers require substantial amounts of ATP to power nitrogen fixation, but are at risk of destroying the nitrogenase if they respire aerobically, the most effective means to generate ATP.

Symbiosis with eukaryotes is one way in which the energy demands of nitrogen fixation can be met. Most nitrogen-fixing symbionts derive most or all of their energy requirements from the host. Of particular note is the association between rhizobia and leguminous plants. As a consequence of the structural organization of the legume nodule and presence of the oxygen-transporting protein, leghaemoglobin, the rhizobia can both fix nitrogen and respire aerobically at high rates in an environment of low free oxygen tension. This remarkable association is described in Chapter 3.

Most nitrogen-fixing symbioses are in plants, but a few animals, fungi, and protists also utilize nitrogen fixation (Table 2.5). The associations in these various groups of hosts are considered here.

2.2.1 Plant symbioses

Three types of nitrogen-fixing bacteria are well known in plants: the rhizobia, actinomycetes of the genus *Frankia*, and cyanobacteria of the family Nostocaceae.

Table 2.5 Nitrogen-fixing symbioses

	Host	Location of symbiont	Symbiont
Plants	Legumes	Root nodules	Rhizobia (*Rhizobium Bradyrhizobium*, and *Azorhizobium*)
	Various dicots[1]	Root nodules (= actinorhizas)	*Frankia*
	Gunnera many cycads *Azolla* (water fern)	Glands at leaf base lateral roots cavities in dorsal lobe of leaves	cyanobacteria (usually *Nostoc*)
	various liverworts and hornworts	Cavities in thallus	
Animals	Termites	Lumen of hindgut	various enteric bacteria, e.g. *Enterobacter*, *Citrobacter*
	Teredinid molluscs (shipworms)	Gland of Deshayes	*Teredinibacter turneri*
Fungi	Lichenized fungi[2]	Cephalodia[3]	Cyanobacteria (usually *Nostoc* or *Scytonema*)
Protists	Marine diatoms: *Rhizosolenia* and *Hemiaulus*	Between cell wall and cell membrane of diatom cell	Cyanobacteria e.g. *Richelia*

[1] Members of eight families: Betulaceae, Casuarinaceae, Coriariaceae, Dastiscaceae, Eleaginaceae, Myricaceae, Rhamnaceae, Rosaceae.
[2] Only 10% of lichenized fungi contain cyanobacteria (most bear algal symbionts; see Table 2.2). Cyanobacterial N fixation is significant in lichens containing both algal and cyanobacterial symbionts (see text for details).
[3] Structures on lichen thallus containing packets of cyanobacteria.

Rhizobia are defined as nitrogen-fixing bacteria that induce, and then inhabit, nodules on the roots of legumes (Fig. 2.7). They have been assigned to two groups in the alpha-Proteobacteria: the genus *Rhizobium*, closely related to the plant pathogen *Agrobacterium*; and *Bradyrhizobium* and *Azorhizobium*, which are allied with *Rhodopseudomonas*. Rhizobia form associations with 17 500 legume species, and also with three species of *Parasponia* (Ulmaceae). Rhizobia cannot nodulate an estimated 12 per cent of legumes, most of which are in the primitive subfamily Caesalpinioideae (Sprent and Sprent 1990). The symbiosis appears to be an advanced feature of legumes and to have evolved independently many times within the Leguminosae (Young and Johnston 1989).

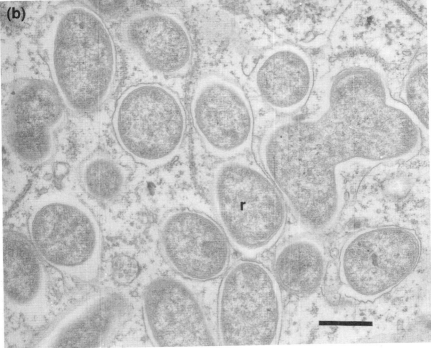

Fig. 2.7 The legume nodule. (a) Nodules on roots of pea plants. (b) Symbiotic rhizobia (r) in an infected cell of a pea nodule. Scale: 0.6 μm. (Micrographs: N.J. Brewin)

The actinomycete *Frankia* is borne in nodules, commonly known as actinorhizas, on the root system of nearly 200 species of nonleguminous dicots, most of which are woody perennials (see footnote 1 to Table 2.5). The vegetative hyphae of *Frankia* extend between the cells of the root cortex. They generate multicellular vesicles, which are the usual site of nitrogen fixation.

Nitrogen-fixing cyanobacteria are known in *c.* 150 species of vascular plants: water-ferns of the genus *Azolla* (6–8 species); most, perhaps all, cycads (*c.* 100 species); and the angiosperm genus *Gunnera* (40–50 species). They also occur on the surface and within the thallus of some bryophytes, but the distribution and abundance of these associations are poorly documented. The cyanobacteria in cycads are in the lateral roots, located between cells of the root cortex. They are also extracellular in bryophytes and *Azolla*, but in *Gunnera* they are intracellular, in gland cells at the base of each leaf (Fig. 2.8). All the cyanobacterial symbionts of plants belong to the family Nostocaceae. The symbiotic growth form is filamentous and comprises vegetative cells, capable of oxygenic photosynthesis, and hetercysts, which fix nitrogen. The symbionts in all plants except *Azolla* are culturable, and most are species of *Nostoc*.

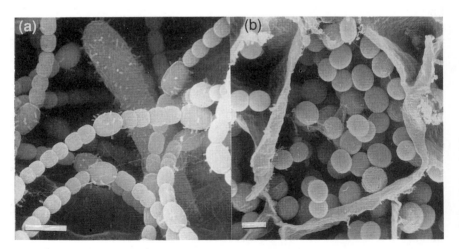

Fig. 2.8 Nitrogen-fixing cyanobacteria in plants. (a) Filaments of the cyanobacterium *Anabaena azollae* in a cavity of the water fern *Azolla*. Scale: 10 μm. (b) Intracellular cells of the cyanobacterium *Nostoc* in gland cells of *Gunnera magallenica*. Scale: 5 μm. (Scanning electron micrographs: R. Honegger)

2.2.2 Animal symbioses

Among animals, nitrogen-fixing symbioses have been demonstrated conclusively only in termites. They may also occur in shipworms (teredinid molluscs) (Table 2.5). Their rarity in animals is not surprising. Animals generate ammonia at substantial rates by the deamination of amino acids, and expend considerable energy to dispose of this potentially toxic compound. To most animals, nitrogen fixation represents a costly way of synthesizing a waste product. Wood-feeding animals, which include many termites and the shipworms, are, arguably, exceptions to this generality because the nitrogen:carbon ratio of wood can be very low.

The nitrogen-fixing symbionts in termites are members of the gut microbiota in the anaerobic hindgut. They can be cultured readily and have been assigned to known taxa, e.g. *Enterobacter agglomerans, Citrobacter freundii*. The significance of these bacteria to the nitrogen nutrition of the insect varies widely between species and with environmental conditions, including diet and season. For example, microbial nitrogen fixation provides for a doubling, within one year, of the nitrogen content of wood-feeding *Nasutitermes* and *Coptotermes* colonies; but nitrogen fixation rates are low in soil-feeding *Rhychotermes* and are barely detectable in the fungus-growing Macrotermitinae (Breznak 1982).

The shipworms are worm-shaped bivalve molluscs which feed on wood fragments generated as they bore into timber. They contain bacteria of the genus *Teredinibacter* (Distel 1990) within an outpocketing of the oesophagus, known as the gland of Deshayes. *Teredinibacter*, when isolated into culture, can both fix nitrogen and degrade cellulose (Waterbury *et al.* 1983). The symbiosis appears to have resolved the major nutritional problems for animals feeding on wood, namely low N:C content with most of the carbon as cellulose. However, the rates of nitrogen fixation and cellulolysis by *Teredinibacter* in the symbiosis are unknown, and the transfer of fixed nitrogen from the bacteria to the animal tissues has not been demonstrated.

2.2.3 Fungal symbioses

The only known nitrogen-fixing symbionts in fungi are the cyanobacteria in lichens (see Tables 2.2 and 2.5). In lichens with cyanobacteria as the sole symbionts, the cyanobacteria have relatively few heterocyst cells (which fix nitrogen) and low nitrogen fixation rates. However, approximately 550 lichen species have both algal symbionts, which are photosynthetic and are distributed throughout the thallus, and cyanobacterial symbionts restricted to special structures called cephalodia. The cyanobacteria in these associations fix nitrogen at high rates, while the photosynthetic requirements of the lichen are met primarily by the algal symbionts.

2.2.4 **Protist symbioses**

The only known nitrogen-fixing symbioses in protists are those between
cyanobacteria and marine, planktonic diatoms of the genera *Rhizosolenia*
and *Hemiaulus*. The cyanobacteria are heterocystous, and are located
extracellularly between the diatom cell membrane and cell wall (Villareal
1987, 1991).

2.3 PROVISION OF NUTRIENTS AND NUTRIENT RECYCLING: INSECT SYMBIOSES

Animals have restricted biosynthetic capabilities and, in particular, they
are dependent on an exogenous supply of essential amino acids and vitamins. Some animals live on diets with very low levels of these nutrients,
and are believed to derive a supplementary supply of essential amino acids
or vitamins from symbionts.

Some of the best-known associations are in insects, and they are summarized in Table 2.6. Most of the symbionts are bacteria, and many
are intracellular, located in specialized insect cells called mycetocytes
(Fig. 2.9). Other microorganisms are located in the insect gut. For
example, the hindgut of cockroaches has a variety of anaerobic bacteria,
and caeca of the midgut in anobiid beetles contain dense populations of the
yeast *Torulopsis*.

The biosynthetic capabilities gained by the insect is known for some of
the associations listed in Table 2.6 (see Douglas 1989), and they are
divisible into three broad groups:

1. Vitamin synthesis. Vertebrate blood is deficient in B vitamins. All
   insect groups that feed in vertebrate blood through the life cycle have
   microbial symbionts that are believed to provide B vitamins. Insect
   groups, such as fleas and mosquitoes, which utilize blood solely as
   adults, lack microbial symbionts.
2. Sterol synthesis. Insects (unlike most other eukaryotes) cannot synthesize
   sterols. Various yeasts, especially in certain planthoppers and the anobiid
   beetles, have been proposed to supplement the low sterol content of
   many plant tissues.
3. Essential amino acid synthesis. This capability may be particularly important in plant sap-feeding insects, such as aphids. The nitrogen in
   plant sap comprises free amino acids, and is dominated by nonessential
   amino acids, which account for 80 mol per cent or more of the total
   amino acid content. (Animals generally require a diet of *c.* 50 mol per
   cent essential amino acids.) The dietary shortfall in plant sap is made
   good by the symbiotic bacteria.

Allied to essential amino acid synthesis is the process of nitrogen recycling. Nitrogen recycling in symbiosis entails symbiont utilization of host nitrogenous waste products, such as ammonia or urea, in the synthesis of 'high-value' nutrients (e.g. essential amino acids), which are made available to the host. In addition to providing nutrients, nitrogen recycling can increase the overall efficiency with which nitrogen is utilized by the association. This can enable animals to utilize low-nitrogen diets, whatever the essential amino acid content of the diet. Nitrogen recycling may be an important function of symbionts in some insects and the gut microbiota in herbivorous mammals (see Chapter 4, section 7).

2.4 ASSOCIATIONS BETWEEN ANAEROBIC PROTISTS AND METHANOGENIC BACTERIA

Methanogenic bacteria generate ATP by synthesizing methane, usually from hydrogen and carbon dioxide. The consumption of hydrogen by methanogens can be advantageous to anaerobic eukaryotes because the rates of oxidative reactions, such as glycolysis, in anaerobes are depressed by high hydrogen tensions. In other words, the bacteria act as a sink for hydrogen (or electrons).

The best-studied methanogenic symbioses are in anaerobic ciliates living in anoxic sediments, landfill sites, and the rumen of cattle and other herbivorous mammals (Fenchel and Finlay 1991*b*). For example, some species of *Plagiopyla* and *Metopus* from landfill sites contain intracellular *Methanobacterium formicicum* (Fig. 2.10), while the rumen ciliates of the family Entodiniomorphida have methanogens on their surface. Methanogens have also been demonstrated in the amoeba *Pelomyxa palustris* (van Bruggen et al. 1988).

2.5 DEGRADATION OF CELLULOSE AND OTHER POLYMERS RESISTANT TO THE DIGESTIVE ENZYMES OF ANIMALS

At least half of the biomass of most plant tissues is cell wall material, which comprises cellulose in a matrix of hemicelluloses, pectic substances, and lignin. Much of this material is resistant to the digestive enzymes of many animals.

Some animals, especially herbivorous mammals, have gained the ability to degrade cellulose and hemicelluloses, but not lignin, by symbiosis with microorganisms in their guts. The microorganisms are restricted to an anaerobic portion of the gut, and they degrade plant polymers to fuel their

Table 2.6 Symbioses in insects

Microorganisms

Insect taxon	Location[1]	Taxa	Metabolic capability	Incidence of symbiosis
Blattaria (cockroaches)	i (fat body)	Bacteria (*Blattabacterium*)	N recycling	Universal
Homoptera (planthoppers, whitefly, aphids, etc.)	i (haemocoel)	Various bacteria and yeasts[2]	amino acid synthesis	Nearly universal
Heteroptera				
Cimicidae (bedbugs)	i (haemocoel)	Coccoid bacteria	B vitamin synthesis	Nearly universal
Triatominae (kissing bugs)	Midgut	Various, includes *Nocardia rhodnii*	B vitamin synthesis	Universal
Various phytophagous bugs[3]	Midgut	Various	?	Widespread
Isoptera[4] (termites)	Hindgut	Various bacteria	N recycling, N fixation	Universal
Anoplura (sucking lice)	i (variable location)	Various bacteria	B vitamin synthesis	Nearly universal
Mallophaga (biting lice)	i (haemocoel)	Various bacteria	?	Irregular
Diptera				
Glossinidae (tsetse flies) Hippoboscidae	i (midgut epithelium)	Bacteria	B vitamin synthesis	Universal

	Location[1]	Microorganism	Function	Distribution
Coleoptera Anobiidae	i (midgut caeca)	Yeasts	Essential amino acid and B vitamin synthesis	Universal
Bostrychidae	i (haemocoel)	Variable	Essential amino acid and B vitamin synthesis	Universal
Cerambycidae	i (midgut caeca)	Yeasts[2]	Essential amino acid and B vitamin synthesis	Widespread
Chrysomelidae	i (midgut caeca)	Coccoid bacteria	Essential amino acid and B vitamin synthesis	Irregular
Curculionidae (weevils)	i (variable location)	Various bacteria	Essential amino acid and B vitamin synthesis	Widespread
Silvanidae (lat grain beetles)	i (haemocoel)	Filamentous bacteria	?	*Oryzaephilus* only
Hymenoptera Formicidae (ants)	i (midgut epithelium)	Bacteria	?	In all Camponoti, irregular in Formicinae

---

[1] i, intracellular; the insect cells containing microorganisms are called *mycetocytes*. For example, the bacteria in cockroaches are borne in mycetocytes, located in the insect fat body.

[2] Yeasts in planthoppers additionally provide sterols.

[3] Coreidae (leaf-footed bugs), Lygaidae (seed sucking bugs), Pentatomidae (stink bugs), Pyrrhocoridae (fire bugs).

[4] 'Lower' termites also contain hypermastigote protists which degrade cellulose.

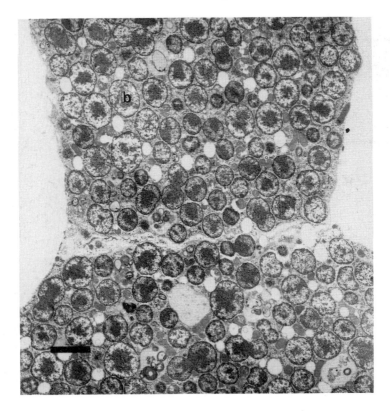

Fig. 2.9 Intracellular bacteria (mycetocyte symbionts of the genus *Buchnera* in the black bean aphid *Aphis fabae*. The bacteria (b) are located in the cytoplasm of the insect mycetocytes. Scale: 5 μm. (Transmission electron micrograph: A. E. Douglas)

own growth. The chief waste products of microbial anaerobic respiration are short-chain fatty acids (SCFAs), such as acetic acid, and these SCFAs are utilized as a source of energy by the aerobic tissues of the animal. If the microorganisms were to degrade cellulose aerobically to carbon dioxide and water, the animal would derive no nourishment.

The synthesis of SCFAs is not a special adaptation to the symbiosis. SCFAs are the waste products of anaerobic respiration by many micro-organisms utilizing a range of carbon sources; the SCFAs are analogous to ethanol produced by fermenting yeasts and lactic acid generated by hypoxic vertebrate muscle. For example, the microbiota in the large intestine of humans produces substantial amounts of SCFAs, contributing up to 10 per

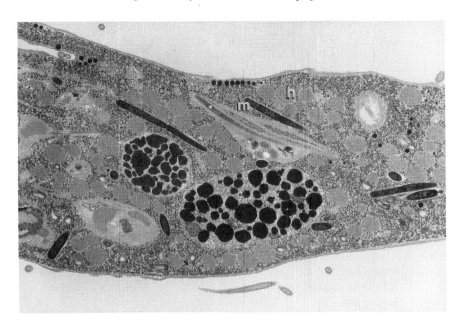

Fig. 2.10 Methanogens in the anaerobic ciliate *Metopus*. The methanogenic bacteria (m) are densely staining elongate rods, located in the cytoplasm of the ciliate host. They are closely associated with hydrogenosomes (h), ciliate organelles which generate hydrogen. Scale: 1.5 μm. (Reproduced from Finlay and Fenchel (1991). *FEMS Microbial Ecology*, **85**, 169–80)

cent of our energy needs, but the chief carbon sources utilized by these microorganisms are carbohydrates in mucus and sloughed cells from proximal regions of the gut (McNeil 1984). Microbial celluloysis is of no significance in the human gut.

The distribution of microbial cellulose degradation in vertebrates is shown in Table 2.7. Virtually all herbivorous mammals and several herbivorous birds and lizards have cellulose-degrading bacteria in the colon, caecum, or both. (This is known as the postgastric cellulolysis because it is distal to the digestive portion of the gut.) Pregastric cellulose degradation in the foregut has evolved at least once in birds (the leaf-eating hoatzin: Grajal *et al.* 1989) and five times in mammals: in the kangaroos, the colobine monkeys, the sloths, the hippos, and in the ancestor of the camels and ruminants (deer, cattle, sheep, etc.) (Bauchop 1977).

Cellulose-degrading microorganisms have also been isolated from the guts of a few insects, but the only association of certain importance to the insect is in a few termites (Chapter 4, section 5).

Table 2.7 Herbivorous vertebrates with cellulolytic microorganisms

	Location of microorganisms in gut		
	Postgastric		Pregastric
	Colon	Caecum	
Eutherian mammals			
Artiodactyla (hippos, camels, ruminants)[1]	+		+
Perissodactyla (horses, rhinos, tapirs)	+	+	
Proboscidea (elephants)	+		
Sirenia (manatees, dugong)	+		
Lagomorpha (rabbits)		+	
Rodentia		+	
Edentata: sloths			+
Primates: Colobinae	+		+
Allouatta (howler monkeys)		+	
Lemuridae		+	
Marsupial mammals			
Macropodidae (kangaroos)	+	+	+
Phascolarctos cinereus (koala)		+	
Vombatidae (wombats)	+		
Pseudocheiridae (ringtailed possums)		+	
Birds			
Tetraonidae (grouse)		+	
Ratites (rhea, ostrich)	+	+	
Hoatzin			+
Reptiles			
Iguanidae	+		
Agamidae	+		
Chelonia (turtles)	+		

[1] The foregut symbiosis probably evolved independently in the hippopotamus and ancestor of camels and ruminants.

2.6 LUMINESCENCE SYMBIOSES

Many organisms generate light by the oxidation of a substrate, generically known as luciferin, with molecular oxygen, in a reaction catalysed by the enzyme luciferase. At least five biochemically distinct luciferin–luciferase systems are known, and luminescence is widespread among eukaryotes, including many protists, fungi, and invertebrate animals, but it is apparently absent from plants and tetrapod vertebrates. Only four genera of bacteria (*Vibrio, Photobacterium, Alteromonas*, and *Xenorhabdus*, all in the gamma-Proteobacteria) are known to include luminescent species.

Most luminescence systems in eukaryotes are intrinsic (i.e. they are not of symbiotic origin). This can be linked to the lack of biochemical barriers to luminescence. Low levels of light are emitted in many oxidative re-actions, and luminescence has probably evolved independently at least 35 times by amplification of such reactions (Hastings 1983).

Bioluminescence symbioses are relatively uncommon, and the lumines-cent symbionts are invariably bacteria. All the hosts are animals: a few nematodes in the terrestrial environment, and some cephalopod molluscs, teleost fish, and the tunicate *Pyrosoma* in the marine environment (Table 2.8). Even within the cephalopods and teleosts, symbiotic light production is less common than intrinsic systems. Nineteen families of cephalopods have luminescent species, but only two of these (Loligidae and Sepiolidae) have bacterial luminescence (Herring 1988); and bacterial light production is known in less than half of the teleost families with luminescent species (McFall-Ngai 1990). The luminescent bacteria in cephalopods and fish are housed in specialized light organs, which are described in the following chapter.

Table 2.8 Luminescence symbioses

Host		Incidence	Symbiont
Order	Family		
(a) *Teleost fish*			
Gadiformes	Macrouridae (grenadiers or rat tails)	146/300 spp.	*Photobacterium phosphoreum*
	Moridae (morid cods)	7/70 spp.	*Photobacterium phosphoreum*
	Merlucciidae (merlucciid hakes)	1/13 spp.	*Photobacterium phosphoreum*
Beryciformes	Trachichthyidae (slimeheads)	6/14	*Photobacterium phosphoreum*
	Anomalopidae (flashlight fish)	5/5 spp.	?
	Monocentidae (pinecone fish)	4/4 spp.	*Vibrio fischeri*
Perciformes	Leignathidae (ponyfish)	20/20 spp.	*Photobacterium leignathii*
	Apogonidae (cardinal fish)	6/? spp.	*Photobacterium leignathii*
Lophiiformes	11 families of Ceratioidei (angler fish)	32/34 genera	?
(b) *Cephalopod molluscs*			
Sepioidea	Sepiolidae		
	Rossiinae	1/3 genera	?
	Heteroteuthinae	5/5 genera	?
	Sepiolinae	4/5 genera	*Vibrio fischeri* in *Sepiola* and *Euprymna*
Teuthoidea	Loliginidae	1/8 genera	*Phosphoreum leiognathii*
(c) *Thaliacean urochordates*		1 genus (*Pyrosoma*)	?
(d) *Nematodes*	Heterorhabditidae	All species	*Xenorhabdus luminescens*

3
Novel structures in symbiosis

Symbiosis has influenced the body form of hosts and (for complex multi-cellular hosts) the anatomy of particular organ systems. The morphological changes linked with symbiosis are usually associated with the housing of the symbionts and enhanced access to the symbionts' metabolic capabilities. The magnitude of the effect varies widely, however, between associations. For example, light capture by the symbiotic algae in marine bivalves of the genus *Tridacna* (the giant clams) is promoted by the asymmetric growth and greatly expanded siphonal tissue of the clam, but the freshwater bivalves *Anodonta* and *Unio* display no substantive morphological modifications linked to the symbiosis with algae.

For some hosts, the morphological consequences of symbiosis are profound. Four instances of morphological innovations are considered in this chapter:

1. Some fish and squid bear luminous bacteria in elaborate light organs, associated with a complex of precisely aligned lenses and mirrors that control light emission.
2. Most herbivorous mammals have greatly enlarged and modified portions of the gut, in which cellulose and other plant macromolecules are fermented by dense populations of microorganisms.
3. The root nodules of many legumes are complex plant organs, which usually develop only in response to compatible rhizobia.
4. Some lichen thalli are among the most complex structures displayed by any ascomycotine fungi.

The last section of the chapter concerns structural novelty at the subcellular level, namely a novel organelle, the symbiosome, which houses most intracellular symbionts.

3.1 THE LUMINESCENCE SYMBIOSES

3.1.1 The light organs

The light organs in animals with luminous bacteria are a mass of narrow tubules, all leading to a single duct which opens to the exterior. The symbiotic bacteria are located in the lumina of the tubules, and they are usually densely packed, for example, at concentrations up to 10^{11} cells ml^{-1} organ fluid in both the squid *Euprymna scolopes* (Ruby and McFall-Ngai 1990) and monocentrid fish (Dunlap 1984). Each bacterium is within a few bacterial cell diameters of the tubule wall (Fig. 3.1(a)). This arrangement ensures that transfer of nutrients and oxygen from the tubule to the

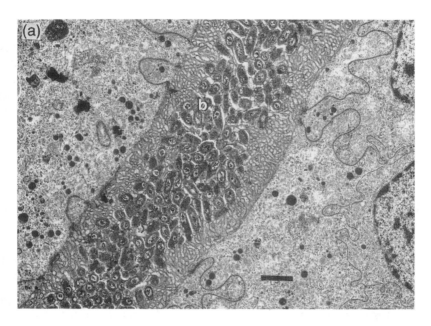

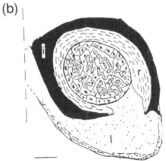

Fig. 3.1 The light organs in the Hawaiian squid *Euprymna scolopes*. (a) A single tubule of the light organ containing a dense population of the luminescent bacterium *Vibrio fischeri* (b). Scale: 3 μm. (Transmission electron micrograph provided by M. McFall-Ngai) (b) Anatomy of the light organ. The tubules bearing the bacteria are central, bounded by a reflector (r), ink sac (i), and lens (l). (Reproduced from Ruby and McFall-Ngai (1992), *Journal of Bacteriology*, **174**, 4885)

bacteria more effectively than could be attained in a single, large cavity containing many bacteria. Additional features which are probably important in maintaining high metabolic activity of the bacteria are that the light organs are well-vascularized and, in some species (including *Euprymna scolopes*), the epithelial cells bear microvilli.

Studies with luminous bacteria in culture suggest that the high density of symbiotic bacteria is essential for light production. In all luminous bacteria, whether or not they are symbiotic, the luciferase is a flavin-linked oxidase comprising two polypeptides, coded by *luxA* and *luxB*, and the expression of these genes is controlled principally by a compound, known as the 'autionducer'. For example, in luminescent strains of *Vibrio fischeri*, the autoinducer is N-(b-ketocaproyl)-homoserine lactone. It is released into the medium, such that its external concentration is proportional to the cell density. At a certain critical concentration of autoinducer, equivalent to approx. 10^6 cells ml^{-1}, luminescence is induced.

3.1.2 The structures that control light emission from the animal

The luminescent bacteria produce light continuously and in all directions, but the value of light production to the animal hosts depends on precise control of both the timing and direction of light emission. Some coastal and mid-water animals produce ventrally directed light, called counter-illumination. This has an anti-predator function. These animals are potentially vulnerable to predators that (from a position below the animal) may detect its silhouette against the downwelling light, and the counter-illumination camouflages the silhouette. Many animals also use light in startle displays, to attract potential prey, and in intra-specific communication.

The emission of light from animals with bacterial luminescence is regulated by animal structures (e.g. muscular shutters and chromatophores) associated with the light organ. These structures tend to be relatively simple in fish with light organs near the animal's surface, but they are often complex in fish with internal light organs and in cephalopods, whose light organs are invariably within the mantle cavity.

Two of the more complex light organ systems, one in the cephalopod *Euprymna*, the other in leiognathid fish, are considered here.

Euprymna scolopes is a small, benthic squid that lives in the coastal waters off Hawaii. It uses light for both counter-illumination and startle displays.

The bilobed light organ is intimately associated with the ink sac in the mantle cavity (Fig. 3.1(b)). Light is emitted equally in every direction from the light organ, and it is reflected back from all sides except the ventral surface by a reflector. The ventral surface of the light organ is covered by a lens, through which light is emitted from the animal. Any light that 'escapes' from the reflector and lens is absorbed by pigment in the ink sac surrounding the reflector. Although the light is invariably ventrally directed, its direction and quality can be altered slightly by small shifts in the position of the reflector.

Ultrastructural studies by McFall-Ngai and Montgomery (1990) indicate that the lens is modified muscle. The reflector comprises cells containing dense arrays of membrane-bound platelets.

Leiognathids are small, silvery fish, common in the Indo-Pacific. They form large, single-species schools in waters of low clarity, and they use luminescence, first as camouflage by counter-illumination, and secondly in communication, both in schooling and in sexual displays.

The light organ of leiognathids is shaped like an American doughnut and encircles the oesophagus of the fish, with the dorsal surface in contact with the anterior end of the fish gas bladder (Fig. 3.2(a)). Light emission from the fish is controlled by three muscles, which act as shutters: two in ventrolateral positions and one between the light organ and gas bladder.

The ventrolateral shutters control the emission of light used in communication. When they are retracted, the light shines through the latero-anterior muscle; when they are closed, no light signal is produced.

The light used in counter-illumination has a more complex passage through the fish body, and the gas bladder plays a crucial role. The gas bladder is lined with purine crystals, which reflect light in all but two regions: the area of contact between the gas bladder and light organ, and the posterior wall. When the shutter between the light organ and gas bladder is retracted, light passes directly into the gas bladder, where it is reflected away from the light organ and dorsal surface, and out, via the posterior 'window', into the hypaxial muscle (Fig. 3.2(b)). This muscle system acts as a light guide and 'diffuser' so that the light which finally emerges through transparent patches in the fish skin breaks the outline of the fish body.

3.2 THE FERMENTATION CHAMBERS IN HERBIVOROUS MAMMALS

3.2.1 Pregastric and postgastric chambers

The breakdown of cellulose, a major component of plant material, is intrinsically a slow process. For a substantial proportion of the cellulose to be degraded in animals, the plant material must be retained in the gut for a

(a)

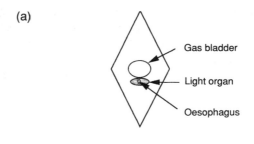

Gas bladder

Light organ

Oesophagus

(b)

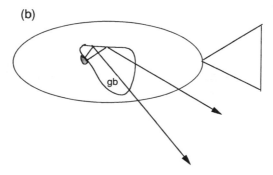

gb

Fig. 3.2 The light organ in leiognathid fish. (a) Diagrammatic vertical section of relationship between light organ (encircling the oesophagus) and gas bladder. (b) Diagrammatic horizontal section showing reflection of light from the dorsal wall of the gas bladder (gb) and out of the fish body, providing counterillumination (see text for details). (Redrawn from McFall-Ngai 1983)

long time (often several days). In herbivorous mammals, cellulose is degraded primarily by microorganisms in enlarged, anaerobic portions of the gut that are known as fermentation chambers.

Virtually all mammalian herbivores have postgastric fermentation chambers in the colon or caecum, or both. Caecal fermentation is characteristic of small mammals, and the colon becomes progressively more important with increasing body size. For example, rodents (<0.1 kg body weight) and rabbits and hares (<5 kg) have greatly expanded caeca (Fig. 3.3(a)), but the colon is important in horses (250–400 kg) (Fig. 3.3(b)) and is the sole site of fermentation in elephants. Caecal fermentation can be linked to the faster passage of food through the alimentary tract of smaller animals. In these groups, food particles of appropriate size for fermentation can be diverted to the caecum, while coarser material is expelled. Consequently, animals with caecal fermentation, but not colonic fermentation, can feed at a faster rate than the transit time of food in the fermentation chamber.

A number of herbivorous mammals additionally have pregastric fermentation chambers (see Table 2.7). Their stomachs are very diverse (Langer 1984). For example, ruminants have three pregastric chambers, in two of which (the rumen and reticulum) fermentation occurs (Fig. 3.3(c)), while kangaroos have a single, elongate fermentation chamber (Fig. 3.3(d)).

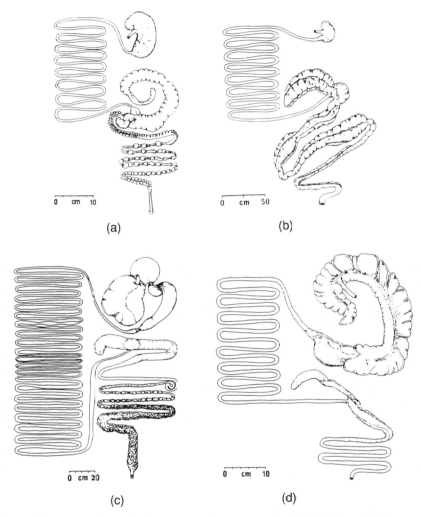

Fig. 3.3 The digestive tracts of herbivorous mammals. (a) Rabbit, with fermentation chamber in the expanded caecum. (b) Zebra, with fermentation chambers in both caecum and colon. (c) Sheep, with foregut fermentation in enlarged portion of the stomach, the rumen and reticulum. (d) Kangaroo, with elongate fermentation chamber in proximal portion of the stomach. (From Stevens (1988))

3.2.2 **Advantages and disadvantages of pregastric fermentation: the theory**

The two principal advantages of pregastric fermentation over postgastric fermentation are:

1. Utilization of recycled nitrogen (see Chapter 2, section 5 and Chapter 4, section 7.3). Any nitrogen recycled by pregastric microorganisms can be absorbed by lower reaches of the animal's gut (especially the intestine), but nitrogen recycled in postgastric chambers would be lost from the animal via the faeces.
2. Detoxification of plant allelochemicals (compounds that deter feeding or are noxious to animals). Ingested plant material is available for detoxification by microorganisms in a pregastric chamber soon after ingestion and before transit through the gut.

For most mammal groups, pregastric fermentation does not enhance the animals' utilization of cellulose to a greater extent than postgastric fermentation. This is because the products of cellulose degradation (short-chain fatty acids) are absorbed efficiently across the walls of all fermentation chambers, whether pre-or post-gastric. In ruminants and camels, however, the utilization of cellulose is enhanced by rumination. This means that the animal regurgitates partially fermented ingesta back to the mouth, where it is chewed again and then swallowed. By subjecting the plant material to repeated cycles of mechanical disruption and fermentation, the cellulose can be broken down very effectively.

The disadvantage of efficient pregastric fermentation is that all the ingested food is metabolized by the microorganisms, whatever its nutritional quality. For some components of food, this results in a reduction in nutritional value. For example, an animal derives more energy from gastric digestion of soluble sugars than from microbial fermentation; fermentation releases very little energy from lipids; and microorganisms transform ingested protein to microbial nitrogenous compounds, approximately 30 per cent of which cannot be digested by the animal.

This disadvantage is minimal for animals that utilize low-quality diets with high cellulose content (and, most particularly, for mammals that ruminate), but it is very severe for small mammals, which are constrained by their high metabolic rate to eat high-quality foods. Dumment and van Soest (1985) have calculated that the critical body weight at which the disadvantage of pregastric fermentation exceeds its value is 3–5 kg. This is the observed body weight of the smallest ruminants (e.g. the dik-dik).

3.2.3 **The gut anatomy and diet of ruminants**

The previous section explains why ruminants and other mammals with pregastric fermentation (especially ruminants) 'should' feed on low-quality, cellulose-rich foods. The problem is that many do not. Almost half of ruminant species, including some large species (e.g. giraffes, moose), feed selectively on easily digestible plant parts. Hofmann (1989) calls these ruminants concentrate-selectors. The rumen of these species is relatively small, and the ingesta pass through it rapidly, such that only half of the ingested cellulose is degraded. Some ruminants feed nonselectively on plant material of high and low nutritional quality, and only a minority feed exclusively on high-fibre foods. In these 'grass/roughage feeders' (which include the domestic cattle, sheep, and goats), approximately 80 per cent of the ingested cellulose is degraded in the rumen.

Postgastric fermentation is more important in concentrate-selector species than in grass/roughage feeders. Hofmann (1989) has estimated that the percentage of ingested cellulose degraded in the caecum/colon is 30–40 per cent for concentrate-selectors, but only 10 per cent for grass/roughage feeders; and the volume ratio of caecum/colon:rumen is 1:8 for concentrate selectors and 1:25 for grass/roughage feeders.

Why do so few ruminant species utilize the potential of pregastric fermentation and rumination to degrade cellulose? One possible explanation is that the pregastric symbiosis evolved in this group under selection pressure for nitrogen recycling and microbial detoxification of plant secondary compounds (see above) and only secondarily acquired a cellulolytic function. Selection pressure for a highly efficient rumen has presumably operated in only a minority of species. These included the ancestors of the domestic cattle, sheep, and goats.

3.3 THE LEGUME NODULE

The legume nodule is a morphological novelty. It arises from a unique plant meristem, which is usually generated only in the presence of rhizobia. The nodule is also unique at the molecular level, because a number of plant genes, known as *nodulins*, are expressed exclusively in this symbiotic organ.

3.3.1 Nodule structure and the exclusion of molecular oxygen from the nitrogenase of rhizobia

A basic function of the nodule is as an environment that promotes sustained nitrogen fixation by the symbiotic rhizobia. Crucial to this function is that a low oxygen tension is maintained within the nodule. This is because the rhizobial nitrogenase enzyme is inactivated at oxygen concentrations of 1–10 μM. In the symbiosis, the enzyme is protected entirely by the plant.

The concentration of oxygen at the centre of the legume nodules is very low ($1 < \mu M$), principally because a zone of cells around the central tissue acts as a barrier to oxygen diffusion. These cells are called the boundary layer. They are very closely packed, usually one to two cells deep and have few air spaces (Fig. 3.4). Consequently, the oxygen diffuses primarily through their cell contents and not through intercellular spaces. The diffusion rate across the boundary layer is very low, simply because the oxygen diffusion coefficient of water is about 10 000 times lower than of air.

Not only is the oxygen concentration is the central tissue of legumes low; it is also independent of the external oxygen concentration. This is because the structural organization of the cortical cells peripheral to the boundary layer varies with external oxygen concentration. As the external concentration increases, the extent of intercellular spaces in the cortex is reduced, and the tissue resistance to oxygen diffusion across the root tissue is increased (Table 3.1).

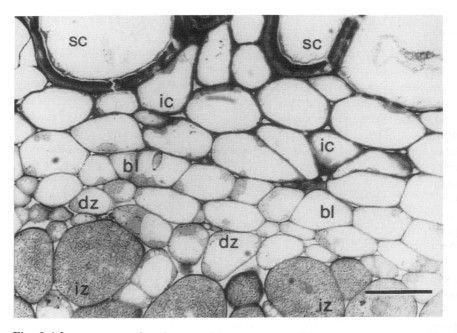

Fig. 3.4 Inner cortex of soybean nodule grown at ambient oxygen tensions. The boundary layer (bl) of thin cortical cells with few intercellular spaces limits the supply of oxygen to the central cells containing rhizobia (infected cells, iz). The extent of the intercellular spaces in other regions of the root cortex (labelled dz and ic) varies with ambient oxygen tension. sc = sclerenchyma. Scale: 40 μm. (From Parsons and Day (1990))

Table 3.1 Structural adaptation of soybean nodules to variation in external oxygen concentration.

The roots of intact soybean plants were maintained at different oxygen tensions for 15 days. The nitrogen fixation rates by rhizobia in the nodules were not influenced by the external oxygen tension, because the passage of oxygen across the cortex was lowered at high oxygen tensions by a reduction in intercellular spaces.

Oxygen pressure (kPa)	Acetylene reduction rate[1] μmol g^{-1} dry weight h^{-1}	Intercellular spaces (% of total area)	
		Outer cortex	Inner cortex
4.17	178	11.0	1.5
19	191	5.3	1.7
47	133	2.0	0.9
75	199	0.7	0.5

[1] Acetylene reduction is a useful index of nitrogen fixation rate (Sprent and Sprent 1990).

(Data from Parsons and Day (1990)).

Despite the very low oxygen concentration in the infected zone of nodules, the bacteria in the infected cells respire aerobically. This is made possible by a plant pigment, leghaemoglobin, in the cytoplasm of the infected cells. Leghaemoglobin binds oxygen with high affinity (the equilibrium dissociation constant is 40–70 nM dissolved oxygen) and so facilitates oxygen diffusion. It has been calculated that at 10 nM free oxygen the concentration of leghaemoglobin-bound oxygen is 70–700 μM, permitting sustained aerobic respiration of the rhizobia (Appleby 1984). An unambiguous demonstration of the importance of leghaemoglobin to the symbiosis is that rhizobia isolated from nodules and maintained at the oxygen tension in nodules cannot respire aerobically unless leghaemoglobin or an alternative oxygen acceptor is provided.

The dependence of rhizobia on plant structures to protect the nitrogenase from oxygen is most unusual. In the nitrogen-fixing cyanobacteria of plants and lichenized fungi, nitrogen fixation is restricted to heterocyst cells; and the sole site of nitrogen fixation in most strains of *Frankia* is multicellular vesicles (see Chapter 2, section 2.1). Both heterocysts and vesicles have thick, oxygen- impermeable cell walls, enabling these structures to fix nitrogen even under atmospheric oxygen tensions.

3.3.2 **The legume nodule at the molecular level: leghaemoglobin and other nodulins**

The legume nodule is ultimately generated from an ordered pattern of gene expression, regulated according to the position and age of the plant cells in the nodule. The pattern can be considered to arise from a developmental programme for nodule formation. The programme is exclusively of plant origin, and it is usually triggered by rhizobia. A few legumes produce nodules in the absence of rhizobia. For example, Truchet *et al.* (1989) describe a cultivar of alfalfa which, under axenic conditions, generates nodules at a frequency greater than 10^{-2}.

Some of the genes expressed in nodules are not expressed in uninfected roots. The proteins produced by these genes are called nodulins. Two groups of nodulins are recognized (Nap and Bisseling 1990).

1. Early nodulins (ENOD), which are synthesized during nodule development. Some early nodulins are proteins associated with plant cell walls. For example, ENOD2 is a wall component of cells in the inner cortex, and ENOD12 is located in the wall of infection threads.

2. Late nodulins (NOD), which are produced in the mature nodule, i.e. after onset of rhizobial nitrogen fixation. They are important in the maintenance of the rhizobia and assimilation of rhizobia-derived ammonia. Leghaemoglobin is a late nodulin. Others include transport proteins in the symbiosome membrane (see section 3.5) and a nodule-specific glutamine synthetase.

The genes for most nodulins are allied to other genes in the legume genome. Appleby *et al.* (1988) have suggested that certain genes have duplicated, that the different gene copies have diverged, and that one has become specialized for function in the nodule. For instance, ENOD2 is structurally similar to (and presumably derived from) hydroxyproline-rich proteins called extensins, present in the cell walls of most elongating plant cells. The genes for some nodulins have unique regulatory sequences. As an example, expression of the gene for nodule-specific glutamine synthetase is not regulated by ammonia (as in other plant glutamine synthetase isozymes), but is probably regulated by signals associated with the invading bacteria (Nap and Bisseling 1990).

One late nodulin that has no known parallel with other legume proteins is leghaemoglobin. However, *Parasponia*, the only non-legume known to bear rhizobia, has a single haemoglobin gene, which is expressed both in nodules and uninfected roots. Appleby *et al.* (1988) have suggested that the haemoglobin in legume nodules has diverged substantially from the putative ancestral legume gene expressed in the uninfected root.

In the early literature there was speculation that the haemoglobin in legumes might be derived from animals by lateral gene transfer. This is erroneous. It is now clear that the legume haemoglobin has an exclusively plant ancestry (Nap and Bisseling 1990).

3.4 THE LICHEN SYMBIOSIS

3.4.1 The lichen thallus

The thalli of many lichens are morphologically complex, in two ways. Many have a defined growth form (foliose, fruticose, etc.: see Fig. 2.2), and the thallus of these lichens is usually stratified into distinct layers (Fig. 3.5). At the simplest, these layers are: the outermost cortex of densely packed fungal hyphae; the symbiont layer, in which the symbiont cells are positioned in regular blocks separated by vertically traversing fungal hyphae; and the medulla of loosely interwoven hyphae.

The complexity of these lichen thalli is not mediated by novel hyphal structures, for hyphal aggregations comparable to the cortex or medulla are known in non-lichenized fungi. The lichen thallus is unique in the ordered, spatial organization of the various hyphal systems (Honegger 1992). Furthermore, the lichen thallus is generated only in symbiosis with algae or cyanobacteria. When growing in isolation, the fungi are morphologically unremarkable.

This situation is reminiscent of the legume nodule. As with legumes, a lichenized fungus can be considered as possessing the developmental programme for thallus formation, which is usually expressed only in the presence of the symbiont. We are, however, totally ignorant of the nature of the putative signals in the lichen symbiosis, and this is in contrast to the detailed molecular understanding of legume nodules (see section 3.3.2 and also Chapter 5, section 4). The difference between our understanding of legumes and lichens reflects the difficulties in any experimental study of lichens. Lichens develop very slowly, and only a few species can be synthesized under laboratory conditions. It takes 4–6 weeks to generate legume nodules from the separate plant and rhizobia, but 6 months to 2 years to synthesize a lichen thallus.

In consequence, virtually all the information on lichens comes from observations of intact thalli. Two topics are considered here: the morphology of lichenized fungi bearing different symbionts, and the organization of the symbiont layer in some stratified lichens.

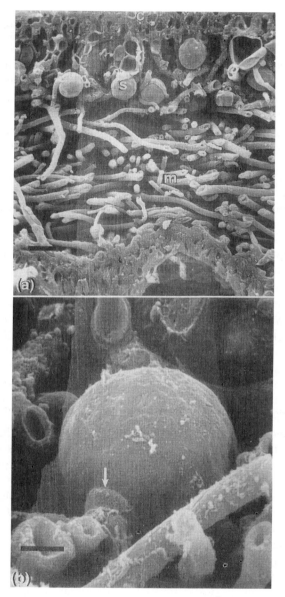

Fig. 3.5 The stratified lichen thallus of *Parmelia borreri*. (a) Vertical section through thallus, showing the algal symbionts (s) sandwiched between two symbiont-free layers, the dorsal cortex (c), and ventral medulla (m). Scale: 10 μm. (b) Morphology of the contact between the fungus and algal symbiont (*Trebouxia*). Each symbiont cell is in contact with a single fungal projection (arrow), known as an intraparietal haustorium. Scale 2 μm. (Scanning electron micrographs: R. Honegger)

3.4.2 The contribution of the fungus and symbionts to the morphology of the lichen thallus

The fungus is traditionally considered as the primary determinant of thallus form. As W. Culberson succinctly stated: 'every different lichen differs because it is the product of a different fungal species' (quoted by Brodo and Richardson 1978). This is consistent with the view expressed above that lichen symbionts make no morphogenetic contribution to thallus structure, but trigger the expression of a pre-existing developmental programme in the fungus. As an illustration, the fungus *Xanthoria parietina* develops into a stratified foliose lichen when it associates with certain *Trebouxia* species, but it forms an amorphous undifferentiated structure when in association with a variety of other *Trebouxia* and *Pleurococcus* species (Ott 1987).

The relationship between lichen symbionts and lichen structure is more complex in a few species. For example, *Sticta filix* can associate with both algae and cyanobacteria (James and Henssen 1976); it has a foliose morphology with green algae, a fruticose morphology with cyanobacteria, and is dimorphic in the presence of both symbionts. A reasonable interpretation of these observations is that *Sticta filix* has two alternative developmental programmes, which are triggered by signals from the different symbionts.

3.4.3 The organization of the symbiont layer in stratified lichens

The algae or cyanobacteria in stratified lichen thalli are positioned in an ordered array (Fig. 3.5). This is believed to promote light capture for symbiont photosynthesis by minimizing shading of one symbiont cell by another.

Let us consider a foliose lichen, such as *Parmelia borreri*. The cells of the algal symbiont *Trebouxia* are maintained in regular rows by the fungus (Honegger 1984). The cell wall of each algal cell is in contact with a single fungal projection, known as an intraparietal haustorium (Fig. 3.5(b)). When *Trebouxia* divides, four daughter cells are produced, and fungal contact is made with each of these daughter cells before the algal mother cell wall is lost (see Fig. 3.6). Then, the fungal haustorium branches and lengthens, so that the symbiont daughter cells become separated and the regular spacing between the algal cells is maintained.

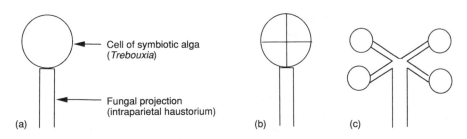

Fig. 3.6 The relationship between projections of lichenized fungi and *Trebouxia* symbionts at symbiont division. (a) Each *Trebouxia* cell is in contact with a single fungal projection. (b) The *Trebouxia* cell divides into four daughter cells. (c) The fungal projection branches into four, each of which elongates, resulting in the separation of the four symbiont daughter cells.

3.5 THE SYMBIOSOME: A NOVEL ORGANELLE IN SYMBIOSIS

The symbiosome is the organelle enclosing intracellular symbionts (Roth *et al*. 1988) (Fig. 1.4). It is generated when symbionts are internalized by endocytosis.

Research on the symbiosome has concentrated largely on the membrane, particularly in legume–rhizobium symbioses. The symbiosome membrane has some features of the cell membrane. For example, in the pea it retains the glycocalyx of carbohydrate, characteristic of the external surface of plant cell membranes (Perotto *et al*. 1991); and in soybean it has a cell membrane-type H^+/ATPase (Blumwald *et al*. 1985). The symbiosome membrane is, however, undoubtedly distinct from the cell membrane. It has a very high ratio of lipid to protein (Robertson *et al*. 1978) and a different profile of proteins from the cell membrane. There is evidence that two sources, additional to the cell membrane, may contribute to the symbiosome membrane in legume nodules. First, some symbiosome membrane proteins are nodulins, i.e. they are unique to the nodule, and absent from all other legume membranes. One example is a dicarboxylate transporter protein (NOD26), which regulates the supply of organic carbon sources to the enclosed rhizobial cell (see Chapter 4, section 2.3). Second, certain symbiont-derived molecules, particularly lipopolysaccharides on the rhizobial cell wall, may become incorporated into the symbiosome membrane (Bradley *et al*. 1985).

Very little is known about the conditions experienced by intracellular symbionts within symbiosomes. In particular, the pH, osmolality, and principal ions within the symbiosome-space are obscure. The presence of

H^+/ATPase in the symbiosome membrane of legumes may be indicative of acidic conditions surrounding the rhizobia, but the symbiosome contents of Cnidaria containing the algae *Chlorella* or *Symbiodinium* are not of low pH (Rands *et al.* 1992).

In many electron micrographs, intracellular symbionts appear to be separated from the symbiosome membrane by a substantial space, representing up to half of the total volume within the membrane. For many symbioses, this is almost certainly a gross overestimate, arising from artefactual shrinkage during tissue preparation for electron microscopy. There is evidence that the symbiosome membrane and enclosed symbiont cells are very closely apposed and may even be in direct molecular contact (Bradley *et al.* 1985).

4

Nutritional interactions in symbiosis

Nutritional interactions are fundamental to most symbioses, because the metabolic capabilities most commonly acquired through symbiosis relate to nutrition. With the exception of a few systems, e.g. luminescence symbioses in animals, the flow of nutrients from the donor to the recipient is substantial. However, the passage of nutrients is usually bidirectional (Fig. 4.1). Intracellular symbionts (which are invariably donors of novel metabolic capabilities) derive their entire nutritional requirement from the surrounding host cell, and many extracellular symbionts are also substantially dependent on host-derived nutrients.

The organisms in symbiosis can derive nutrients from their partners in three ways:

   biotrophy: from living cells of the partner;
   saprotrophy: from dead material;
   necrotrophy: by killing cells of the partner, which are then utilized saprotrophically.

In most associations, nutritional interactions are biotrophic. For example, the photosynthetic compounds that animals, such as corals, derive from their symbiotic algae are released from intact and photosynthetically active cells, and not by lysis of the algae.

Biotrophic nutrient transfer encompasses three topics:

   How much is translocated?
   What are the mobile compounds (i.e. the compounds transferred between the partners)? and
   How is nutrient release induced in the symbiosis?

Biotrophic nutrition has been studied extensively in photosynthetic symbioses (mycorrhizas, lichens, and algal associations in protists and invertebrates). As section 4.1 describes, the basic characteristics of biotrophic nutrient transfer have been established for these systems, but how the symbiosis induces photosynthate release is still obscure. This aspect of biotrophic nutrient transfer is much better understood for nitrogen-fixing systems (see section 4.2).

Sections 4.4–7 concern cellulose degradation, vitamin transfer, and nitrogen recycling, especially in associations between animals and their gut

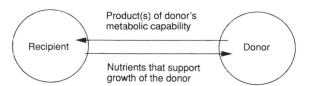

Fig. 4.1 Bidirectional nutrient transfer in symbiosis. In symbioses involving micro-organisms, the donor is usually the symbiont and the recipient is usually the host.

microbiota. These gut symbioses are unusual in that their nutritional relations are not exclusively biotrophic. The microorganisms are partly biotrophic, deriving nutrients of low molecular weight (e.g. nitrogenous compounds) from the host, and partly saprotrophic because they utilize food ingested by the animal host. Likewise, the animal partner gains nutrients from its microbial symbionts by a combination of biotrophy (e.g. by absorption of microbial fermentation products) and necrotrophy (vitamins and essential amino acids are derived primarily from digestion of microbial cells).

4.1 PHOTOSYNTHETIC SYMBIOSES

4.1.1 The amount of photosynthate transferred

The substantial release of photosynthate in photosynthetic symbioses was first established by radiotracer experiments. The symbiosis is incubated with ^{14}C-carbon dioxide, and both the total radioactivity fixed by the association and the amount in the recipient (fungus, protist, or animal) are quantified.

For symbioses with animals and protists, the estimates of photosynthate transfer are very variable, ranging from 10 per cent to 60 per cent of the total carbon fixed by the algae or cyanobacteria. In part, this reflects real differences between associations and, in temperate regions, reduced trans-location rates in winter (Sutton and Hoegh-Guldberg 1990). Published values for associations involving fungi are very consistent: 10–20 per cent for mycorrhizal fungi (Harley and Smith 1983); and 60–80 per cent for the fungi in foliose lichens (Smith 1980). The data for lichens may, however, be misleading because the experiments were conducted with fully saturated thalli incubated on buffer solutions. The proportion of photosynthate translocated in thalli of lower water content may be different, but this has not been investigated.

Radiotracer techniques may cause systematic underestimates of photo-synthate transfer, for two reasons. First, the heterotrophic partner is likely to utilize some of the photosynthate as a respiratory substrate and, compounding this loss, the respired carbon dioxide may be refixed by photosynthesis. Second, the fungal or animal tissue cannot be separated quantitatively from its partners, with the result that much of the ^{14}C translocated to the recipient is scored as retained in the photosynthetic partner.

Muscatine *et al.* (1984) have developed an alternative approach to quantify the transfer of photosynthate, requiring neither radioactivity nor physical separation of the partners. The total oxygen flux (in the light) and respiration rate (in the dark) of the intact symbiosis are measured, and the biomass of the symbionts is determined. From these data, two parameters are calculated: the photosynthetic carbon fixed by the algal cells, and the carbon requirements for algal growth and respiration. The difference between these two parameters is taken as the amount of photosynthate translocated to the host.

To date, the technique of Muscatine *et al.* (1984) has been applied only to marine symbioses between animals and the alga *Symbiodinium* (Muscatine 1990). The calculated values of photosynthate transfer can exceed 90 per cent for shallow-water, tropical symbioses, especially for Cnidaria (e.g. corals and sea anemones). For these systems, it appears that the symbionts are photosynthesizing at substantial rates, and are releasing virtually all their photosynthetic carbon to the host tissues.

4.1.2 **The identity of the mobile photosynthetic compounds**

The algae or cyanobacteria in lichens, animals, and protists release substantial amounts of photosynthate when isolated from the symbiosis, and it is widely accepted that the released compounds are the same as the mobile compounds in the symbiosis. However, the rate of photosynthate release from the isolated symbionts declines, usually to undetectable levels, within minutes to hours of isolation, and symbionts in long-term culture release no more photosynthate than nonsymbiotic algae or cyanobacteria.

The photosynthate released from the isolated symbionts is usually in the form of one or a few compounds of low molecular weight, and the nature of the compounds is determined by the symbiont taxon (Table 4.1). For example, among the lichen symbionts, the cyanobacteria release glucose and the algae release polyols (also known as polyhydric alcohols) (Smith 1980). Most released compounds are not major intracellular compounds in the symbiont cells. As an example, symbiotic *Chlorella* release the sugar maltose and not the principal internal photosynthetic product, sucrose.

There are a few exceptions to these generalizations. *Symbiodinium* may release lipids (of high molecular weight), in addition to compounds of low molecular weight (e.g. glycerol); and the algae in lichens have substantial

Table 4.1 Compounds translocated from photosynthetic symbionts to their hosts

The photosynthate released from most photosynthetic symbionts is almost exclusively in the form of one or a few compounds of low molecular weight, the identity of which is determined by the symbiont and not host.

Host	Symbiont	Released compound(s)
Lichenized fungi	Cyanobactera	
	Nostoc, Scytonema	Glucose
	Algae	
	Trebouxia, Coccomyxa	Ribitol[1]
	Stichococcus	Sorbitol
	Trentepohlia	Erythritol
Protists and invertebrates	Cyanobacteria	Glucose
	Algae	
	Chlorella	Maltose
	Symbiodinium	Glycerol, triglyceride

[1] Ribitol, sorbitol, and erythritol are polyols.

intracellular reserves of the mobile polyol. (The polyol destined for release is not, however, derived from the major intracellular pool, but from a small, distinct pool of newly synthesized polyol.)

The chemical identity of compounds translocated to mycorrhizal fungi is uncertain. Smith and Smith (1990) suggest that the fungi may utilize glucose and fructose. These monosaccharides would be derived from the hydrolysis of sucrose, the principal photosynthetic sugar transported to the root in the phloem sap.

4.1.3 The mechanisms underlying photosynthate transport

A central feature of photosynthate transfer is its requirement for intimate contact between the partners of the association. In mycorrhizas, the fungus probably obtains photosynthate only via specialized structures, such as the arbuscules in VA-mycorrhizas and the Hartig net in ectomycorrhizas (see Chapter 2, section 1.3 for details); and, as described above, photosynthate release by most algae and cyanobacteria declines rapidly on isolation from the association. The fungus or animal partner is believed to modify the metabolism and transport properties of the photosynthetic partner in such a way as to promote synthesis and release of the mobile compounds. The mechanisms are unknown, but there are three alternative hypotheses accounting for this effect: cell-wall degradation, release factors, and metabolic imbalance.

1. *Cell-wall degradation.* The mobile compounds are thought to be either components of the photosynthetic partner's cell wall, or cell-wall precursors diverted from wall synthesis to the heterotrophic partner. This hypothesis has been developed independently for lichens, mycorrhizas, and alga–invertebrate symbioses, but there is persuasive contrary evidence. For lichens and alga–invertebrate symbioses, the mobile products of algae and cyanobacteria are not allied metabolically to their cell-wall components (Table 4.1). In mycorrhizas, the plant cell walls apposing arbuscules and the Hartig net of mycorrhizal fungi are often amorphous in structure, but this is almost certainly a response to the pressure exerted by fungi on the plant cell, and not (as once believed) a consequence of fungal breakdown of the plant cell wall.

2. *Release factors.* It is suggested that a specific compound (known as a release factor) is produced by the host, and that it triggers photosynthate release from algal symbionts. Release factors have been implicated in associations between *Symbiodinium* and marine invertebrates because homogenates of some (but not all) hosts stimulate photosynthate release from isolated *Symbiodinium* (Trench 1971; Sutton and Hoegh-Guldberg 1990). No symbiont other than *Symbiodinium* responds to host homogenate, but isolated *Chlorella* release photosynthate when incubated in acidic conditions (Cernichiari *et al.* 1969), indicating that low pH may be the release factor for *Chlorella*. However, the biochemical bases of the effects of host-homogenates and low pH are unclear, and it is uncertain whether these effects with isolated cells are relevant to nutrient release in the intact association. In particular, the symbiosome contents enclosing *Chlorella* are not markedly acidic (Rands *et al.* 1992).

3. *Metabolic imbalance.* Rees and Ellard (1989) have argued that one necessary condition for sustained photosynthate release is an imbalance between the carbon and nitrogen metabolism of the photosynthetic symbionts. Specifically, the symbionts may have poor access to nitrogen, and therefore have an excess of carbon compounds derived from photosynthesis. The excess may be released to the host. This hypothesis is relevant to symbiotic algae and cyanobacteria (but not to mycorrhizas; mycorrhizal plants are rarely nitrogen-limited). There are three lines of supportive evidence: the photosynthate released by most symbionts does

not contain nitrogen (see Table 4.1); the growth of algal symbionts in animals is probably nitrogen-limited (see Chapter 6, section 2.1); and in the coral symbioses, the proportion of photosynthate translocated to the host declines when the sea water is supplemented with nitrogen (McCloskey, unpublished results).

4.1.4 **The uptake of photosynthetic compounds by the heterotrophic partner**

In the intact symbioses, the mobile photosynthetic compounds are taken up across the cell membrane of the mycorrhizal and lichen fungi, and across the symbiosome membrane or cell membrane of protists and animals. A proportion of photosynthate is respired, and the remainder is metabolized very rapidly to other compounds. These metabolic transformations render the carbon unavailable, by backflow, to the photosynthetic partner.

The fate of photosynthate is known, at least in outline, for most associations. The lichenized fungi transform much of the photosynthate to their own storage polyols (mostly mannitol and arabitol). The VA-mycorrhizal fungi convert the plant photosynthate to lipid and glycogen, and the other mycorrhizal fungi synthesize trehalose and mannitol. Among animals, the Cnidaria convert the photosynthate from *Chlorella* to glycogen, and that of *Symbiodinium* into lipid.

4.1.5 **The significance of the photosynthate to the recipient organism**

(a) *Mycorrhizal fungi*

The mycorrhizal fungi are nutritionally dependent on the symbiosis, deriving most or all of their carbon requirements from the plant. Most ectomycorrhizal fungi have very limited capacity to utilize the complex organic macromolecules in soil (these are the principal carbon sources for saprotrophic fungi), and VA-mycorrhizal fungi can grow only in association with plants.

(b) *Animals and protists*

In most animals and protists, the photosynthate provided by the symbiosis is supplemented by holozoic feeding or dissolved organic matter absorbed from the ambient water, or both. Two complementary approaches have been adopted to investigate the relative importance of algal photosynthate and these alternative sources of nutrients.

The first approach is to examine the response of the symbiosis to continuous darkness. Frost and Williamson (1980) have studied a natural population of the freshwater sponge *Spongilla lacustris* containing *Chlorella* (Fig. 4.2). Over 7 weeks, sponges shaded by black plastic grew to less than half the size of sponges under natural illumination, but enclosure of sponges to prevent feeding did not significantly depress their growth rates. These data indicate that the algal symbionts make a major contribution to the growth of *S. lacustris*. The growth of many scleractinian coral symbioses in their natural habitat is also dependent on illumination. In the 12-month study of Wellington (1982), shading dramatically reduced the growth rates of both *Pocillopora damicornis*, a shallow-water branching species, and the deeper-water and more massive corals of the genus *Pavonia*.

The second approach is to determine directly the contribution of photosynthetic carbon to the total carbon utilized by the animal in respiration. This can be done using parameters obtained to quantify photosynthate transfer by the method of Muscatine *et al.* (1984) (see section 4.1.1). Some shallow-water Cnidaria derive sufficient photosynthate to fuel their entire respiratory needs, but deep-water species may derive only part of their carbon and energy requirements from their algal symbionts (Muscatine 1990).

(c) *Lichens*

Lichens have very little control over their water content and, under some environmental conditions, they experience several cycles of drying and wetting in a single day. Most of the photosynthate that the fungus derives from its symbionts is retained as polyols and is utilized in lichen tolerance of desiccation and rapid wetting.

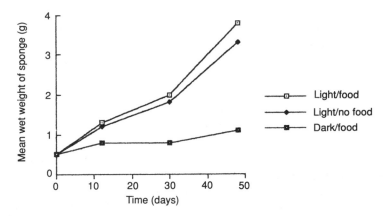

Fig. 4.2 Growth response of the sponge *Spongilla lacustris* to illumination and food availability. (Redrawn from data in Frost and Williamson (1980))

The fungal polyols are important in lichen thalli of low water content, for two reasons. First, they decrease the water potential of the lichen, and this facilitates absorption of water from the air and reduces water loss. Second, their hydroxyl groups can replace 'bound' water associated with proteins and other macromolecules, so preventing denaturation in thalli with very low water contents.

Polyols also contribute to the fungal response to rapid wetting. Over the first 0.5–2 minutes, substantial amounts of solutes are lost from the fungal cells, and subsequently the lichen respires at very high rates (known as resaturation respiration) for up to 2 hours. The symbiont-derived polyol pool provides virtually all the carbon lost by these processes. Farrar (1976) has argued that the utilization of polyols, and not polysaccharide reserves or other components of the fungus, enables the lichen to respond to frequent changes in hydration without perturbing its metabolism. This effect, known as physiological buffering, is dependent on the continuous supply of carbon from the photosynthetic symbionts.

4.2 NITROGEN-FIXING SYMBIOSES

4.2.1 Transport of nitrogen fixation products from symbiont to host

A substantial proportion of the nitrogen fixed by the nitrogen-fixing symbionts of plants and lichens is released to the host. This has been demonstrated for rhizobia, *Frankia*, and cyanobacterial symbionts. Virtually nothing is known of nitrogen translocation in symbioses with animals and protists.

The mobile compound in most nitrogen-fixing symbioses in plants and lichens is ammonia, the product of the nitrogen fixation reaction. However, the cyanobacteria in cycads may metabolize ammonia to glutamine or citrulline prior to release to the host, and the mobile compound(s) in *Frankia* associations are uncertain (Pate 1989).

Most symbionts release ammonia probably because their capacity to utilize this compound is repressed in symbiosis. In the symbiotic cyanobacteria, the activity of the ammonia-assimilating enzyme, glutamine synthetase, is very low or undetectable in symbiosis (Table 4.2); and in the cyanobacteria–*Azolla* symbiosis, transcription of the cyanobacterial glutamine synthetase gene is repressed (Nierzwicki-Bauer and Haselkorn 1986).

Suppression of the symbionts' capacity to utilize the product(s) of their own metabolic capability may be a general mechanism mediating nutrient transfer in nitrogen-fixing symbioses. It cannot, however, be applied to other symbioses. In particular, there is no evidence that photosynthate

Table 4.2 Enzymes of ammonia assimilation in the lichen *Peltigera aphthosa*

The cyanobacterial symbiont *Nostoc* assimilates nitrogen via glutamine synthetase, and the fungus via glutamate dehydrogenase. In the cephalodia (the portion of the lichen thallus containing *Nostoc*), enzyme activity is predominantly fungal glutamate dehydrogenase.

| | Enzyme activity (μmol product min^{-1} mg^{-1} protein) | |
	Glutamine synthetase	Glutamate dehydrogenase
Nostoc in culture	60	2
Medulla of lichen thallus (contains no symbionts)	0	24
Cephalodia of lichen thallus	2	398

(Data from Rai *et al.* (1981)).

transfer (considered in section 4.1) is linked directly to the suppression of the symbionts' capacity to assimilate photosynthetic carbon.

In most nitrogen-fixing associations, the ammonia is probably transferred to the host tissues by simple diffusion, with a steep ammonia concentration gradient between symbionts and host (Udvardi and Day 1990). For example, in soybean nodules, the concentration of ammonia is 12 mM in the rhizobial cells and undetectable (<10 μM) in the cytoplasm of the infected cell (Streeter 1989). The gradient is maintained by the sustained production of ammonia in the symbiont cells and its rapid assimilation in the host tissues, which have very high activities of ammonia-assimilating enzymes (glutamine synthetase–glutamate synthetase in plants and glutamate dehydrogenase in lichenized fungi: see Table 4.2).

4.2.2 *nif* gene expression in symbiotic bacteria

If the nitrogen-fixing genes (*nif* genes) of rhizobia were regulated as in nonsymbiotic bacteria, such as *Klebsiella pneumoniae*, rhizobia would not fix nitrogen in symbiosis. This is because the *nif* genes of *Klebsiella* are repressed by ammonia, via the *ntr* system (as outlined in Fig. 4.3(a)). In rhizobia, however, expression of the nitrogen-fixation genes (*nif* and *fix*) is regulated by oxygen, and not nitrogen (Fig. 4.3(b)). For example, when *Rhizobium meliloti* is maintained at low oxygen tensions, a membrane-

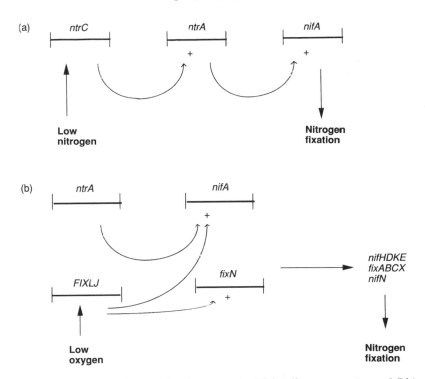

Fig. 4.3 Regulation of nitrogen fixation genes in *Klebsiella pneumoniae* and *Rhizobium meliloti* (a) *Klebsiella pneumoniae* fixes nitrogen when combined nitrogen sources are in short supply. The key regulatory gene is *nifA*, whose product (NifA) is a transcriptional activator of all other *nif* genes. The transcription of *nifA* requires a sigma factor (a component of RNA polymerase) coded by *ntrA*, which in turn is controlled (via *ntrC*) by the availability of combined nitrogen sources. (b) *Rhizobium meliloti* fix nitrogen under microaerobic conditions, independent of the combined nitrogen concentration. This bacterium has two regulatory genes (*nifA* and *fixN*) controlling the expression of nitrogen fixation genes (*nif* and *fix*). The expression of *nifA* and *fixN* is independent of nitrogen supply, and is regulated by *fixLJ* in response to oxygen tension.

bound haemoprotein (FixL, coded by the gene *fixL*) phosphorylates a regulator protein FixJ, which, in turn, activates the transcription of the regulatory genes, *nifA* and *fixK* (de Philip *et al.* 1990).

The mechanisms underlying the regulation of nitrogen fixation in cyanobacteria and *Frankia* are unknown, but they are presumably linked to the induction of heterocyst and vesicle differentiation, respectively, in the presence of combined nitrogen.

4.2.3 **Carbon sources utilized by nitrogen-fixing symbionts of plants**

All the symbionts of plants require organic compounds from their hosts. This includes the cyanobacteria, which either photosynthesize at very low rates (in *Azolla* and bryophytes) or are incapable of photosynthetic carbon dioxide fixation (in *Gunnera* and cycads).

The supply of organic carbon compounds to nitrogen-fixing symbionts has been studied in most detail in the legume–rhizobium symbiosis. The carbon requirements of rhizobia are met from legume photosynthesis, with sucrose transported to the nodules via the phloem. The nodules utilize between 10 and 30 per cent of total plant photosynthate, the proportion varying with environmental conditions and the age and species of plant. Sucrose is, however, transported very slowly across the symbiosome membrane to rhizobia (Day and Copeland 1991), suggesting that this sugar is not an important carbon source for rhizobia.

The principal carbon compound utilized by symbiotic rhizobia is the dicarboxylic acid, malate, perhaps with a second dicarboxylic acid succinate. The evidence is two-fold: biochemical and genetic.

1. Biochemical. Rhizobia, freshly isolated from nodules, assimilate malate (probably as the univalent anion) at high rates from low external concentrations. Exogenous malate stimulates nitrogen fixation by isolated rhizobia (McDermott *et al*. 1989; Day *et al*. 1990).

2. Genetic. The transporter that mediates malate uptake by the rhizobia is coded by the gene *dct*. *dct* mutants of rhizobia form small nodules, and they cannot fix nitrogen (Finan *et al*. 1983).

The malate utilized by the rhizobia in symbiosis is probably generated from the fermentation of sucrose by plant cells in the central, microaerobic portion of the nodule, as shown in Fig. 4.4.

The legume symbiosome membrane enclosing rhizobia also bears a dicarboxylate transporter. This plant transporter is a nodulin, NOD26 (nodulins are introduced in Chapter 3, section 3.2). In the soybean symbiosis, the K_m for malate transport across the symbiosome membrane is 106 μM, and across the rhizobial membrane it is 9 μM (Udvardi *et al*. 1989), indicating that the symbiosome membrane has a much lower affinity for malate than the enclosed rhizobial cells. These data suggest that the symbiosome membrane is a major control point in the supply of malate to the bacteria.

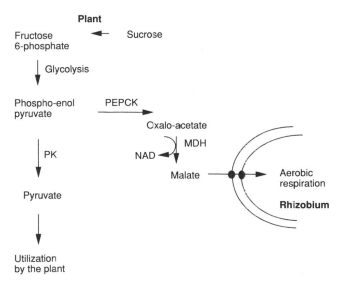

Fig. 4.4 Provision of malate to rhizobia in the legume symbioses. The crucial branch-point in degradation of sucrose in nodules is the fate of phospho-enol-pyruvate (PEP). In nodule cells, oxaloacetate synthesis via PEPCK (PEPcarboxy-kinase) is favoured over pyruvate synthesis via PK (pyruvate kinase), because nodule tissues contain several PEPCK isozymes, including one not expressed in other legume tissues (i.e. a nodulin, see Chapter 3) with high affinity for PEP. Malate synthesis via PEPCK and MDH (malate dehydrogenase) is directly advantageous to plant cell metabolism because NAD, required for sustained glycolysis is regenerated. The malate is taken up by the rhizobia cells and used as a respiratory substrate, to generate ATP for nitrogen fixation.

4.3 MYCORRHIZAL FUNGI AND THE MINERAL NUTRITION OF PLANTS

4.3.1 Mycorrhizal fungi and the acquisition of immobile ions

Phosphate is one of the most immobile soil nutrients required by plants, and soluble phosphate is available at very low concentrations, usually less than one-thousandth of the concentration in plant tissues. Most plant roots have high-affinity uptake systems for phosphate and, in many soils, they assimilate phosphate more rapidly than it diffuses through the soil. As a result, the concentration of phosphate in the immediate environs of the root progressively declines. This phosphate 'depletion zone' around the root cannot be overcome by any physiological process of the plant, and sustained phosphate uptake depends on continual growth out of the depletion zone into previously unexplored volumes of soil.

The structures that would maximize phosphate uptake would be fast-growing and of small diameter, to maintain a high surface area/volume relationship for absorption. Plant root hairs and hyphae of mycorrhizal fungi, with a diameter <10 μm, are much more effective structures for phosphate absorption than roots (whose diameters usually exceed 100 μm). In general, the fungal hyphae extend much further from the root surface than root hairs, and they can form branched filamentous structures that explore a larger volume of the soil than the unbranched root hairs.

In summary, mycorrhizal fungi can acquire phosphate from soils at a higher rate that roots, primarily because they can grow rapidly out of developing depletion zones in the soil. This capability results in enhanced incorporation into the root because phosphate is transferred along the hyphae to the root faster than it could diffuse through the soil.

Ammonium is also a relatively immobile ion; and mycorrhizal fungi contribute to its acquisition by plants, in the same way as for phosphate. Fungal uptake of ammonium is particularly important in the ectomycorrhizal coniferous and deciduous forests, where ammonium is the principal nitrogen source in the soil.

4.3.2 Nutrient acquisition in ericoid mycorrhizas

The fungus in ericoid mycorrhizas does not contribute to plant uptake of immobile phosphate and ammonium ions. Most of the phosphorus and nitrogen in the acidic soils bearing ericoid plants is in the form of recalcitrant organic compounds in the humus. The fungus releases phosphatases (which mineralize phosphate) and proteolytic enzymes, and the products, phosphate and amino acids, are absorbed by both the fungus and plant roots (Read 1983).

4.4 HERBIVOROUS MAMMALS AND MICROBIAL DEGRADATION OF CELLULOSE

As described in Chapter 2, section 5, microorganisms in the anaerobic portions of animal guts derive energy from the fermentation of organic compounds, with short-chain fatty acids (SCFAs) as the principal waste products. SCFAs diffuse freely across the gut wall into the tissues of the animal, where they can be utilized as substrates for aerobic respiration. The biotrophic transfer of SCFAs is of particular nutritional significance for herbivorous mammals containing cellulolytic microorganisms.

The fermentation chambers of herbivorous mammals bear a wide taxonomic diversity and very high densities of microorganisms. For example, the rumen of a cow feeding on grass contains: 10^9-10^{10} bacterial cells (200–400 species) ml^{-1}, approximately 10^5 ciliate protists (40–50 species) ml^{-1}, and many chytrid protists (often known as gut-fungi). The biomass of each of these three groups of microorganisms is approximately equal. Some of the bacteria and ciliates and all chytrids are cellulolytic.

The dominant cellulolytic bacteria in ruminant mammals are *Ruminococcus albus*, *R. flavefaciens*, and *Fibrobacter* (=*Bacteroides*) *succinogenes*. Their cellulases are beta-1,4-endoglucanases, and the major end-products of cellulolysis are succinate and SCFAs, especially acetate. *Ruminococcus* and *Fibrobacter* have also been isolated from the caecum of cattle, sheep, and horses. The cellulolytic bacteria in wild herbivores have not been studied extensively, but are assumed to be similar.

Chytrids produce extracellular cellulases (both exo-and endo-glucanases) and hemicellulases. The principal degradation products are acetate, formate, lactate, carbon dioxide, and hydrogen (Pearse and Bauchop 1985). The motile phase of the chytrid life cycle (known as the zoospore) preferentially colonizes fibrous plant fragments, particularly vascular tissue, and subsequently generates rhizoids, which penetrate into the plant material (Fig. 4.5(a)). Chytrids are important in the degradation of fibrous fragments because their rhizoids can weaken the structure of the plant material and make new surfaces available for colonization by cellulolytic bacteria. They are present in mammals utilizing low-quality, high-fibre diets, such cattle, kangaroos, horses, and elephants (Bauchop 1989), but absent from small species (such as rabbits) feeding on high-quality plant material.

Among the ciliate protists, the capacity to degrade cellulose is restricted to the larger and structurally complex members of the order Entodiniomorphida. Examples in the rumen of cattle include *Epidinium*, which adheres to plant fragments by pseudopodia and releases cellulases, and *Polyplastron*, which engulfs intact plant fibres (Fig. 4.5(b)). Ciliates are responsible for up to 25 per cent of cellulolysis in domestic ruminants feeding on grass, but if they are experimentally eliminated the animal is not affected deleteriously (Bird and Lang 1978).

In summary, the bacteria are crucial to cellulolysis in all mammalian herbivore symbioses. The ciliate and chytrid protists contribute to cellulose breakdown in animals on high-fibre diets, but neither of these two groups of protists is believed to be essential.

4.5 CELLULOSE DEGRADATION IN INSECTS

Insects that feed on living plant material, such as leaves, green shoots, and young roots, are called *phytophages*. These insects do not degrade dietary

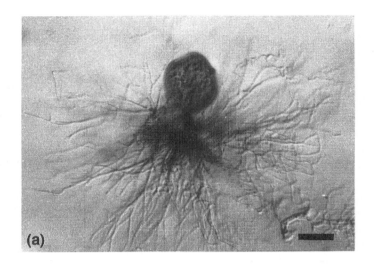

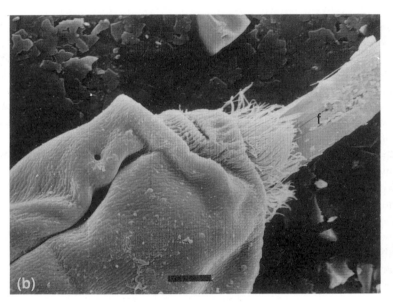

Fig. 4.5 Cellulolytic microorganisms in herbivorous mammals. (a) The chytrid protist *Neocallimastix* with rhizoids extending into the substratum. Scale: 40 μm. (Micrograph: I. Heath) (b) The ciliate protist *Polyplastron* ingesting a fragment of plant fibre (f). Scale: 6 μm. (Scanning electron micrograph from Bohatier *et al*. 1990))

cellulose, even though the cellulose represents about half the biomass of their food. Some phytophages, such as caterpillars, ingest large amounts of plant material, utilize the easily assimilated nutrients, and void the cullulose in their frass. Other groups, including the Orthoptera (grasshoppers, locusts, etc.) possess cellulose-degrading bacteria, but the microbiota does not contribute to their nutrition (Charnley *et al.* 1985).

Virtually all the carbon in wood is in the form of lignocellulose, and most wood-feeding (xylophagous) insects have cellulases. The best-studied xylophages are the termites. All termites have intrinsic cellulases in the midgut (Hogan *et al.* 1988*a*, *b*) (Table 4.3). In the family Termitidae (the 'higher' termites, comprising 75 per cent of all termite species), these enzymes are the sole source of cellulolysis, but other termite families (collectively known as the 'lower' termites) additionally have cellulose-fermenting protists in their hindguts (Table 4.3 and Fig. 4.6). The protists are trichomonads and hypermastigotes. They are obligate anaerobes, and they ferment cellulose to acetate. If lower termites are deprived of these symbionts, they die (Yamin and Trager 1979).

Cellulases of fungal origin are utilized by fungus-growing termites (Termitidae of the subfamily Macrotermitinae). These termites maintain the fungus *Termitomyces* in their communal nests. The fungal mycelium grows on sponge-like structures (combs) prepared by the termites. The fungus produces aggregations of fungal conidiophores, known as nodules, on which the termites feed. Fungal cellulases derived from the nodules remain active in the termite gut (Martin 1991), but the quantitative significance of the fungal enzymes is uncertain. Veivers *et al.* (1991) have found that fungal-derived cellulases contribute less than 0.03 per cent of the total cellulase activity in *Macrotermes michaelsoni*.

A minority of cockroaches are xylophages. One of these woodroaches, *Cryptocercus*, utilizes cellulolytic protists closely similar to the symbionts in

Table 4.3 Sites of cellulase activity in termites (Data from Veivers *et al.* (1982) and Hogan *et al.* (1988*a*))

Gut region	Enzyme activity against crystalline cellulose (% of total activity)	
	Nasutitermes walkeri (a 'higher' termite)	*Mastotermes darwiniensis* (a 'lower' termite)
Foregut and salivary glands	5	13
Midgut	94	13
Hindgut	1	74

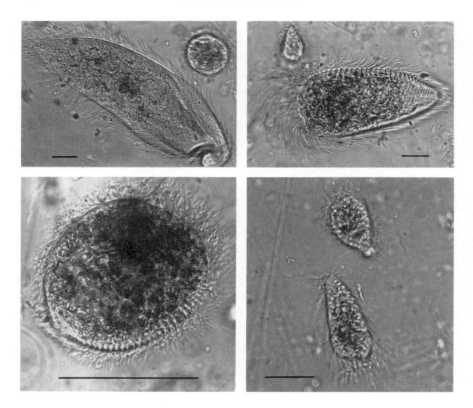

Fig. 4.6 Cellulolytic protists in the lower termite *Coptotermes lacteus*. Scale bar: 25 μm. (Micrographs: M. Slaytor)

lower termites, but a second, *Panesthia*, probably produces intrinsic cellulases (Scrivener *et al.* 1989). Many other xylophagous insects have cellulase activity in their midgut. These include beetles (cerambycids, anobiids, and bupestrids) and larvae of siricid wasps. These systems are not based on anaerobic, cellulolytic gut microorganisms, and the relative importance of intrinsic cellulases and ingested fungal enzymes is uncertain (Martin 1991).

In summary, very few insects derive carbon and energy from cellulolytic microorganisms in their guts. The sole properly documented association is in the lower termites (accounting for 550 species), which obtain SCFAs from cellulose-degrading protists in their hindguts.

Why do so few insects utilizing cellulose-rich diets have a cellulolytic microbiota? Several factors may be important.

1. The growth and population increase of phytophagous insects are limited by dietary nitrogen (not carbon), and therefore cellulose degradation

would almost certainly not improve their performance. Cellulolysis is only likely to be important in xylophagous insects because virtually all the carbon in wood is in the form of lignocellulose. Mammals have a much greater demand for carbon than insects because of their very high metabolic rate, linked to the maintenance of a high, uniform body temperature. This is probably why herbivorous mammals need to utilize dietary cellulose.

2. Some insects, but no vertebrates, can synthesize cellulases. Insects can therefore potentially degrade cellulose independently of microbial symbionts.

3. Insects are small and most fly. They cannot readily support a capacious gut bearing slowly fermenting plant material. Significantly, most individuals of lower termites (with a cellulolytic gut microbiota) do not fly.

4.6 SYMBIOSIS AS A SOURCE OF VITAMINS

The usual approach to investigate symbiont provision of vitamins is dietary. The requirement of a host for a particular vitamin is established, both in the presence and absence of symbionts. If the performance of the host lacking symbionts varies with dietary concentration of the vitamin, but that of hosts with symbionts is independent of the dietary vitamin supply, then the symbionts may represent an important source of that nutrient.

Symbiont provision of vitamins has been studied in two groups of animal: blood-sucking arthropods and mammals, especially herbivores.

4.6.1 **Blood-sucking arthropods**

All arthropods that feed on blood through the life cycle have intracellular bacterial symbionts. These include ticks and several insects, e.g. the sucking-lice (Anoplura), bedbugs (Cimicidae), and tsetse-flies (of the Diptera Pupiparia) (Douglas 1989; see Table 2.6). For each of these groups, elimination of the microorganisms leads to impaired growth and fertility, which can be partially restored by supplementing the blood diet with B-vitamins. The implication is that the symbionts provide B-vitamins, which are in short supply in blood. It is not known whether the arthropod hosts derive the vitamins from intact symbiont cells (biotrophy) or by lysis of the symbionts (necrotrophy).

4.6.2 **Mammals**

Mammals derive various vitamins from members of their gut microbiota, primarily by necrotrophy. Release of vitamins from living microbial cells is probably not of nutritional importance to the host, although it cannot be discounted entirely.

The groups best suited to necrotrophic vitamin acquisition are those with pregastric fermentation, such as the ruminants. Cattle, for example, derive their entire vitamin-B and -K requirements by degradation of microorganisms that pass from the rumen to the gastric portion of the digestive tract.

Mammals lacking a substantial pregastric symbiosis may acquire vitamins by necrotrophy in two ways.

1. Microbial cells inhabiting the small intestine are digested; this may be an important source of vitamin B_{12} in humans (Albert *et al.* 1980).
2. The animal may gain access to the microbiota in postgastric chambers (the caecum or colon; see Fig. 2.11) by feeding on faeces enriched with the microorganisms or microbial products. This behaviour is termed coprophagy.

Coprophagy is a regular part of the behaviour of rabbits, rodents, shrews, and phalangerid marsupials, and it is occasionally observed in other mammals, especially if they are maintained on inadequate diets. There has been very little study of the precise nutritional value of coprophagy to the animal, beyond the slow weight gain by rabbits and rats prevented from feeding on their faeces. Rats (but not rabbits) restrained in this way develop specific dietary requirements for vitamin K and vitamin B_{12}, indicating the importance of coprophagy to the vitamin nutrition of these rodents.

4.7 NITROGEN CONSERVATION AND RECYCLING

4.7.1 The nitrogenous waste products of animals

The processes of nitrogen conservation and recycling arise directly from the metabolic limitations of animals. Animals have no general means of storing nitrogen (in particular, glycogen and lipid do not contain nitrogen). Amino acids and other nitrogenous compounds in excess of an animal's immediate requirements are therefore degraded, for example by gluconeogenesis to glucose, with the production of ammonia. Ammonia is toxic and must be eliminated either directly or after transformation to other compounds, such as urea or uric acid.

4.7.2 Nitrogen conservation

Many microorganisms, including some symbionts in animals, can utilize

ammonia, urea, or uric acid as nitrogen sources. Let us consider a microbial symbiont that derives its entire nitrogen requirements from its animal
host. Many animals are nitrogen-limited (this is because animals are mostly
protein), and the symbionts' nitrogen needs could therefore impair animal
performance. This cost of symbiosis would be reduced, and even eliminated, if the symbionts used the animal's nitrogenous wastes.

Symbiont utilization of their host's nitrogenous wastes is known as
nitrogen conservation (Fig. 4.7(a)). This process is indicated by increased
net production of nitrogenous waste compounds when the symbionts are
eliminated, and it probably occurs in many animal symbioses. The best-
characterized example is the utilization of urea by the gut microbiota of
mammals. Urea passes by diffusion from the bloodstream directly into the
gut lumen, where it is metabolized to ammonia by urease enzymes (mostly
of microbial origin) and then assimilated, mostly into bacterial protein and
cell walls. In humans, about 25 per cent of the circulating urea is utilized by
the hindgut microbiota.

4.7.3 The distinction between nitrogen conservation and nitrogen recycling

The principal advantage of nitrogen conservation to animals is that it
reduces the costs associated with nourishing the symbionts and eliminating
nitrogenous waste compounds. The nutritional advantage to the animal
depends on the fate of the nitrogen in the microorganisms. If the nitrogen
is channelled exclusively into symbiont maintenance and growth, the animal
gains nothing in a nutritional sense. This is probably the situation with the
uricolytic bacteria in the hindgut of termites. However, the ammonia, or
an allied compound, could be used to synthesize compounds of considerable nutritional value to the animal, such as essential amino acids, and then
release them to the host. This is nitrogen recycling (Fig. 4.7(b)).

Nitrogen recycling probably occurs in the mycetocyte symbioses of both
cockroaches and aphids. The intracellular bacteria in both groups of insects
utilize their hosts' waste nitrogen and provide essential amino acids to the
insect tissues (Douglas 1989; Whitehead *et al.* 1992), but the incorporation
of host-derived nitrogen into amino acids released to the insect tissues has
yet to be demonstrated directly.

Nitrogen recycling is also widely assumed to occur in associations between Cnidaria, especially corals, and algae. The biochemical evidence is,
however far from persuasive, and two aspects of the association are not
readily compatible with nitrogen recycling. First, the photosynthetic compounds released by the algal symbiont *Symbiodinium* contain no nitrogen
(see Table 4.1). Second, the growth of algal symbionts is probably
nitrogen-limited (see Chapter 6, section 2.1), suggesting that the flux of
nitrogenous compounds from the host to symbionts is not substantial.

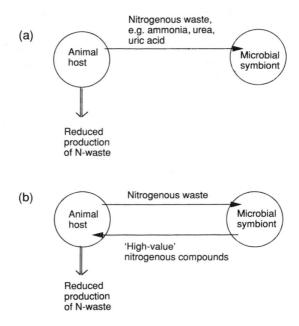

Fig. 4.7 Nitrogen conservation and recycling. The net production of nitrogenous waste products by an animal host is reduced when the symbionts utilize the compounds as a nitrogen souce. (a) The symbionts *conserve* nitrogen if they utilize the host-derived nitrogen solely to support their own growth. (b) The symbionts *recycle* nitrogen if they transform host-derived nitrogen to compounds of nutritional value to the animal (e.g. essential amino acids), which are then translocated to the host.

The one system in which nitrogen recycling undoubtedly occurs is the gut symbioses in herbivorous mammals, and this is considered below.

4.7.4 Nitrogen recycling by necrotrophy in mammalian gut symbioses

A substantial part of the nitrogen assimilated by the gut microbiota of mammals is incorporated into microbial protein. Ruminants and other mammals with foregut symbioses recover the protein by digesting the microorganisms every time that a portion of the rumen contents is transferred to the gastric region of the stomach.

The value of nitrogen recycling in ruminants depends on the composition of microorganisms in the rumen, especially the ratio of bacteria to protists.

If the protists are eliminated experimentally from sheep, the flow of non-ammonia nitrogen to gastric digestion is increased from an average of 2.43 to 2.85 g per 100 g organic matter (Veira 1986). Protists engulf bacteria and ingested protein, but (unlike the smaller bacteria) they are retained in the rumen. As a result, the protists are effectively competing with the animal for protein. This can be deleterious to the animal, especially for young individuals on low-nitrogen diets.

Necrotrophic recycling of nitrogen cannot occur in mammals with a hindgut microbiota, unless the animal exhibits coprophagy (section 4.6.2). Otherwise, the nitrogen is conserved, but not recycled. Nitrogen recycling by coprophagy probably occurs in rabbits (Thacker and Brandt 1955), but it has not been investigated in other mammals.

5

How symbioses are formed

Research on the formation of symbioses includes four topics.

1. Identification of the source of the partner. A host may acquire its symbionts either from the environment or directly from another host. In many associations, the symbionts are transferred directly from a host-parent to its offspring, and this process is known as vertical transmission.
2. Establishment of the symbiosis. It is usual to consider the development of an association as a series of stages, each stage dependent on the successful completion of the previous stage. As a hypothetical example, an association between an animal host and microbial symbiont may include contact, internalization of the microorganism, and initiation of nutrient transfer.
3. Specificity of the association. This refers to the taxonomic range of partners with which an organism can form a symbiosis. Specificity is a consequence of both the degree of specialization of an organism for its partner, and its capacity to select and discriminate between alternative potential partners.
4. Recognition. In the context of symbiosis, recognition is the means by which certain taxa are selected and others are rejected, i.e. the mechanisms underlying the specificity of a symbiosis. (The term 'recognition' as applied to symbiotic interactions has a different meaning from its use in cell biology to describe highly specific molecular interactions between cells.)

5.1 MODES OF FORMATION OF SYMBIOSES

5.1.1 Animals

Symbionts are usually transmitted vertically from parent to offspring in asexually reproducing animals (especially Cnidaria and sponges), but the source of symbionts acquired by sexually produced animals is variable. All intracellular symbionts in insects are vertically transmitted; virtually all luminescent and chemosynthetic bacteria in marine animals are gained from freeliving populations in the water column; and the source of gut symbionts and algal symbionts varies between associations. The associations described below illustrate this diversity.

(a) *Vertically transmitted mycetocyte symbionts in insects*

The mycetocyte symbionts in insects are transferred from the mycetocytes of females to each egg developing in the ovaries (Buchner 1965). After fertilization of the egg, the symbionts are retained within the developing embryo, and are finally allocated to the embryo's mycetocytes as these cells differentiate.

Eberle and McLean (1983) have investigated the transmission of the mycetocyte symbionts in the human body louse, *Pediculus humanus*. The mycetocytes are aggregated together as a coherent organ, the mycetome, just ventral to the midgut. During the final moult of the insect, the bacteria are expelled from the mycetome of females (Fig. 5.1(a)), and they migrate to the reproductive tract. The migration is rapid and is usually complete within 12 hours. In adult female lice, all the bacteria are consequently associated with the reproductive organs. (The bacteria are retained within the mycetome of adult males.)

The bacteria in adult females are initially associated with the lateral oviducts (Fig. 5.1(b)) and not their final destination, the insect's ovaries. The bacteria penetrate the oviduct wall and enter the insect cells lining the oviduct (Fig. 5.1(c)). The transfer of bacteria from the oviduct to the ovaries has not been studied by modern techniques but, according to the

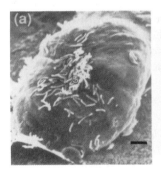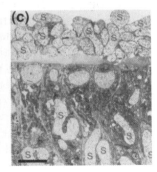

Fig. 5.1 Vertical transmission of mycetocyte symbionts in the human body louse *Pediculus humanus*. (a) The rod-shaped bacterial symbionts (S) emerging from a single opening of the female mycetome. Scanning electron micrograph. Scale: 10 μm. (b) The symbiotic bacteria (S) accumulate on the lateral oviduct of the louse. Scanning electron micrograph. Scale: 10 μm. (c) The bacteria (S) are taken into insect cells lining the oviduct. Transmission electron micrograph. Scale: 5 μm. (Reproduced from Eberle and McLean (1983))

study of Ries (as summarized in Buchner (1965), the bacteria 'pass' to the insect cells at the junction between each oviduct and the base of the ovary (a region known as the pedicel). In *P. humanus*, as in all insects, the ovary comprises a number of ovarioles, each containing a string of eggs at different developmental stages. (The most mature egg is at the base of the ovariole, abutting the pedicel.) A sample of bacteria is transferred from the pedicel into the cytoplasm of the basal egg in each ovariole, just before the egg is released into the oviduct. In this way, each egg is inoculated in turn with its mother's bacteria.

(b) *Luminescence symbioses*
The light organs of marine fish and cephalopod molluscs are colonized by free-living bacteria.

The acquisition of luminescent bacteria has been described in the squid *Euprymna scolopes*, which has symbionts of *Vibrio fischeri* in a bilobed light organ in the mantle cavity (McFall-Ngai and Ruby 1991). (The structural organization of the light organ and associated complex of mirrors and shutters is described in Chapter 3, section 1). Water is passed through the mantle cavity of *E. scolopes* for gas exchange and locomotion, and it travels over the light organ, which, in symbiont-free animals, has a pair of epithelial flaps (Fig. 5.2(a)). Water-borne bacteria are trapped by cilia and microvilli on these flaps and are transported to pores, providing entry to the light organ. Once in the light organ, the bacteria divide rapidly and, within a few days, the organ is luminescent. Simultaneously, the epithelial flaps regress (Fig. 5.2(b)).

(c) *Invertebrate symbioses with photosynthetic algae and cyanobacteria*
For these associations, the source of symbionts varies with host taxon. The flatworms and bivalve molluscs acquire their symbionts from the free-living environment; the cyanobacteria in marine sponges are transmitted vertically via the egg; and in the Cnidaria both vertical transmission and acquisition from the environment are widespread.

The algal symbionts obtained from the environment by Cnidaria and other invertebrate animals are acquired by the feeding route. Contact between the partners may be aided by water currents, generated by the animal for feeding, and also by behavioural attributes of the algae. For example, motile cells of the dinoflagellate *Symbiodinium* exhibit chemotaxis towards alga-free Cnidaria, and flagellate cells of *Tretraselmis* accumulate around egg capsules of their flatworm host *Convoluta roscoffensis*. Many host species ingest any potential symbionts that enter their feeding currents. The Cnidaria, however, are, carnivores, and very few cnidarian hosts feed directly on algal cells. (One species that does is the scyphistoma stage of the jellyfish *Cassiopea*.) Most Cnidaria probably acquire sym-

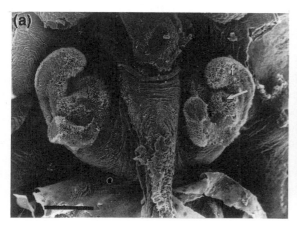

Fig. 5.2 Morphological changes in the light organ of *Euprymna scolopes* associated with the acquisition of luminescent symbionts. (a) The light organ of a 1-day-old juvenile bearing epithelial flaps (f), which intercept potential symbionts in the water current, and pores (arrow). Scale: 120 μm. (b) Light organ of 18-day-old animal, which has been infected with luminescent bacteria. Note regression of the epithelial flaps. Scanning electron micrograph. Scale 120 μm. (Micrographs: M. McFall-Ngai).

bionts by feeding on herbivorous zooplankters with viable algal prey in their guts.

5.1.2 Plants

Most symbioses in plants are formed at each sexual generation of the plant. This applies to all root symbioses (including the relationships with rhizobia, *Frankia*, mycorrhizal fungi, and cyanobacteria in cycads) and several shoot-borne associations, e.g. *Azorhizobium* in stem nodules and cyanobacteria in *Gunnera*. Exceptionally, the cyanobacteria in the water fern *Azolla* are retained through the life cycle.

The fungi and bacteria associated with plant roots are very widely distributed in soils. They have a considerable ability to survive in isolation, either in a metabolically active condition (e.g. rhizobia) or as dormant spores (e.g. VA-mycorrhizal fungi). They can also be dispersed over considerable distances, by wind or animals. Many soils with established vegetation additionally have a substantial hyphal network of mycorrhizal fungi associated with functional mycorrhizas; and, in such soils, these hyphae are probably the usual source of mycorrhizal infection (see Fig. 1.3). It is only in very exceptional circumstances (e.g. soils of extreme pH, temperature, or salinity, and new substrata, generated for example by volcanic eruptions) that soils lack bacterial or fungal propagules.

Exudates from plant roots may promote directed growth of mycorrhizal fungi or proliferation of rhizobia in the soil. These effects have been interpreted as evidence for specific plant signals, but it has proved very difficult to distinguish between nonspecific nutrient-rich conditions in the rhizosphere and symbiosis-specific interactions. For example, one volatile signal believed in early studies to promote germ-tube growth of VA-fungi has been shown to be carbon dioxide, generated by root respiration (Becard and Piche 1989).

The response of roots to contact with rhizobia is immediate and pronounced. In many legumes, the rhizobia infect root hairs, which then curl and branch. Thereafter, two processes are initiated (Fig. 5.3):

1. *Invasion.* The plant cell wall is stimulated to invaginate and grows inwards, to form an apoplastic 'tunnel', known as the infection thread. The infection thread, with its enclosed rhizobia, passes across the plant cell walls and through cells, and it may branch frequently. In many legumes, especially herbaceous species, contact is made with a suitable plant cell and the rhizobia are endocytosed; a cell containing rhizobia is called an infected cell. However, in many tropical trees, the rhizobia are retained in the infection thread, and are not endocytosed (Sprent and Sprent 1990).

2. Cortical cell division. Within 12–24 h of rhizobial–root contact, cells in the underlying root cortex start to divide rapidly. These rapidly proliferating cells form the meristem that gives rise to the nodule tissue.

Plant roots make no discernible response when contacted by compatible mycorrhizal fungi. In particular, the plant does not mount a defensive response (e.g. deposition of lignin or callose, production of phenolics), as is observed in interactions with pathogenic fungi. The hyphae of mycorrhizal fungi extend between plant root cells, primarily by mechanical means (Marks and Foster 1973). The fungi may also secrete enzymes, such as pectinases and cellulases, but this has not been demonstrated.

5.1.3 Lichens

The source of symbionts acquired by lichen fungi is closely correlated with the mode of reproduction of the fungal host. Virtually all sexually produced spores of the fungus are dispersed without symbionts, but most asexual propagules bear symbiont cells derived from the parental thallus. An example of asexual dispersal structures is the soredium, which comprises a loose ball of fungal hyphae enclosing a group of symbiont cells.

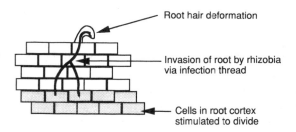

Root hair deformation

Invasion of root by rhizobia
via infection thread

Cells in root cortex
stimulated to divide

Fig. 5.3 Response of legume root to invasion by rhizobia.

Some lichens rarely (and possibly never) reproduce sexually. These asexual species are assured of compatible symbionts, and this enables them to colonize substrata very rapidly. In consequence, they have a considerably wider distribution and higher abundance than most sexual species. Poelt (1970) has recognized 'species-pairs' in many genera of lichenized fungi, comprising a sexual 'primary' species and a more common asexual 'secondary' species. The primary species of some asexual lichens are unknown, and may be extinct.

The spores of lichen fungi acquire symbionts either from the environment or another symbiosis. Both sources are believed to be utilized, but their relative importance is uncertain. Aggressive interactions between lichen fungi are known. For example, a widespread lichen *Xanthoria parietina* invariably reproduces by spores, which, on germination, acquire *Trebouxia* cells from soredia of another common lichen, *Physcia tenella* (Ott 1987). There are also a number of fungi, called lichenicolous fungi, that infect established lichens, kill the fungal partner, and take over the symbionts.

Lichenologists have for many years attempted to synthesize lichens from the separate fungal and symbiont partners. Using low nutrient conditions, often with slow wet–dry cycles, a few lichens (e.g. *Cladonia cristatella*, *Usnea strigosa*, and *Xanthoria parietina*) can now be synthesized, but most species tested develop only to an amorphous aggregate of fungal hyphae and symbiont cells. The reasons for their failure to develop further are unknown.

5.1.4 Protists

When protist hosts reproduce asexually, their intracellular symbionts are usually transmitted vertically by simple partition between the daughter cells. The life cycles of many protists are, however, complex and poorly understood, and it is often unclear whether symbionts persist through the full life cycle.

The fate of symbionts at sexual reproduction of protists is particularly uncertain. It is, however, known that *Chlorella* symbionts are transferred between conjugating pairs of the ciliate *Paramecium bursaria*, at very low frequency. Chesnik and Cox (1987) have demonstrated that the nuclei of the chrysophyte symbionts in gametes of the dinoflagellate *Peridinium balticum* fuse soon after fusion of their host cell nuclei, indicating that at least this stage in the sexual cycles of host and symbiont is synchronized.

5.1.5 Conclusions

Vertical transmission of symbionts is advantageous to a host, in that the host is assured of gaining a compatible symbiont (or, at least, a symbiont compatible with its parent). Vertical transmission is the norm at asexual reproduction of hosts, whether mediated by fragmentation, binary fission, or specialized asexual propagules (as in some lichenized fungi), and it is widespread, but not universal, in sexually reproducing hosts. Two factors may limit the incidence of vertical transmission: structural barriers in the host and costs of vertical transmission.

(a) *Structural barriers in the host*

Structural barriers in the host may restrict access of symbionts to the host's gametes. For example, the symbionts in lichens do not usually occur in the lichen cortex, where the fungal spores are produced; and symbionts of plant roots are physically distant from the seeds borne on the plant-shoot system. The gut wall of animals represents a major barrier to microorganisms, and contact between the gut microbiota and host offspring can be made only by contamination of the egg surface or the neonate. (Intracellular symbionts of animals are often transmitted via the egg cytoplasm.)

(b) *The costs of vertical transmission*

Vertically transmitted symbionts are supported through early development of the host, when they may not be required. The costs are twofold: space, and nutrients. For example, the mycetocyte symbionts in insects occupy up to 10 per cent of the egg volume, and they often proliferate rapidly after transfer to the egg, presumably utilizing the egg's nutritional reserves.

The costs of vertical transmission may be particularly important for marine invertebrate hosts with larval stages in their life cycle (e.g. many Cnidaria and bivalve molluscs). All bivalve hosts of algae (e.g. *Tridacna*) and chemosynthetic bacteria acquire their symbionts from the sea water, after metamorphosis to the adult form, and many sexually reproducing Cnidaria also acquire their algal symbionts from the environment.

Potentially there is also a long-term cost of vertical transmission. Because the host obtains its complement of symbionts from a parent, it lacks access to a variety of alternative symbiont taxa. As yet, the importance of this cost has not been considered in detail. Vertically transmitted symbionts are however, unlikely to evolve characteristics detrimental to their host, for they are 'trapped' within their host lineage and the well-being of their host is to their advantage. (For sexually reproducing hosts, this applies only to symbionts transmitted through one parent. If transmitted via both egg and sperm, the symbiont would have a twofold advantage over host genes, and could spread even if detrimental to the host. This is presumably why symbionts are transmitted vertically through one sex, usually the female.)

5.2 SPECIFICITY AND TAXONOMY

A firm grasp of the taxonomy of the interacting organisms in symbiosis is a prerequisite for any investigation of specificity. Until recently, this has been lacking because many microbial symbionts cannot be classified by traditional classification schemes. However, as will be described in section 5.2.2, symbiotic systems are amenable to molecular methods in microbial systematics. The coming years should witness dramatic developments in our understanding of the taxonomic position of microbial symbionts and the specificity of symbioses.

5.2.1 Difficulties in accommodating symbiotic microorganisms in traditional microbial systematics

Most traditional classification schemes require isolation of microorganisms into culture, followed by detailed growth and biochemical studies. Many microbial symbionts, especially intracellular forms, are unculturable, and so cannot be accommodated within established schemes. Unculturable bacterial symbionts (especially in protists and animals) are often described as: 'allied to rickettsias, mycoplasmas, and other unculturable forms'. This interpretation has no foundation, because unculturability is a character of little taxonomic value.

5.2.2 The molecular revolution in microbial systematics

The organization and sequence of DNA are ideal characters for phylogenetic studies because they uniquely reflect the evolutionary history of organisms. Molecular techniques have revolutionized the field of microbial systematics. The relationships between culturable taxa have been reassessed (Woese 1987), and it is now realized that culturable microorganisms represent just a small proportion of the microbial diversity in natural environments (Giovannoni *et al.* 1990).

Molecular techniques are having a considerable impact on studies of specificity in symbioses. It is now possible to classify unculturable symbionts and to investigate the genetic diversity among the symbionts in a single host taxon.

(a) *Identification of unculturable microorganisms*

Ribosomal genes (especially the 16S rDNA of bacteria and 18S rDNA of eukaryotes) are widely used for phylogenetic studies because they are present in all organisms and evolve slowly. Portions of the 16S/18S rDNA of unculturable microorganisms can be amplified by the polymerase chain reaction. The amplified DNA is then sequenced, and the microorganism can be identified by comparisons with sequences of other taxa.

Using these techniques, it has been demonstrated that the unculturable bacteria (mycetocyte symbionts) in various homopteran insects are allied to various well-known bacterial groups. For example, the bacteria in aphids and whitefly belong to distinct lineages in the gamma-Proteobacteria, while the bacteria in mealybugs are beta-Proteobacteria (Munson *et al.* 1991; Clark *et al.* 1992). These data indicate that the mycetocyte symbioses in Homoptera have evolved independently several times.

The 16S rRNA genes in the unculturable chemosynthetic bacteria of bivalve molluscs and Pogonophora have also been investigated (Distel *et al.* 1988). These bacteria form a coherent group within the gamma-Proteobacteria, suggestive of a single evolutionary origin.

Most luminescent symbionts in the light organs of fish and cephalopod molluscs are culturable, and they have been assigned to various species of *Vibrio* and *Photobacterium* (in the gamma-Proteobacteria). The symbionts in a few fish have not, however, been cultured, and molecular studies suggest that these forms are not members of any known species. For example, the symbiont in the Caribbean flashlight fish *Kryptophanaron alfredi* is probably a new species allied to *Vibrio harvei* (Haygood 1990), and the bacteria in the deep-sea angler-fish (Ceratioidae) can be assigned to neither *Photobacterium* nor *Vibrio* (Haygood *et al.* 1992).

(b) *Concordance between the phylogenies of hosts and symbionts*

As part of their study on the bacterial symbionts of aphids, Munson *et al.* (1991) analysed the 16S rDNA sequence from the bacteria in 11 aphid species. From these data, they constructed a preliminary phylogeny of the bacteria, and found that the relationships between the bacteria closely mirrored the phylogeny of their aphid hosts (Fig. 5.4).

The phylogeny of the aphid symbionts parallels their hosts because the bacteria are vertically transmitted. The association probably evolved in the common ancestor of the Aphidoidea during the Permian (some 250 million years ago), and the bacteria have been passed from mother to offspring ever since. As the Aphidoidea diversified, each aphid lineage carried its own complement of bacteria with it. As a consequence, the symbionts diversified with the aphids.

The genetic relationships among hosts and symbionts have been compared in a few other vertically transmitted systems, including the cockroaches and their mycetocyte symbionts (Wren *et al.* 1989) and the water-ferns *Azolla* and cyanobacteria (Plazinski *et al.* 1988). In both these systems, the host and symbiont phylogenies are closely matched.

If organisms can select their partners, for example when hosts acquire symbionts from the environment, the host and symbiont lineages will be

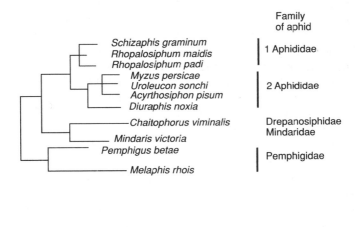

25 base changes

Fig. 5.4 Phylogenetic tree of *Buchnera*, the mycetocyte symbiont of aphids based on 16S rDNA sequence. The relationship between the 11 symbiont isolates is concordant with the phylogeny of their aphid hosts. 1, tribe Aphidini; 2 tribe Macrosiphini. (Redrawn from Munson *et al.* (1991))

interchanged and mixed, and a close correlation between host and symbiont phylogenies would not be expected. This prediction is borne out for
associations between marine Cnidaria and algae of the genus *Symbiodinium*.
Genetic variation has been detected between *Symbiodinium* isolates, but
this variation does not reflect Cnidarian phylogeny. Using molecular techniques, Rowan and Powers (1991) have identified 10 *Symbiodinium* taxa in
22 cnidarian species; some closely related Cnidaria bore different symbionts, and the *Symbiodinium* in certain taxonomically divergent hosts
were indistinguishable.

5.2.3 Symbiosis as a taxonomic character

Host range and the capacity to form an association are used as taxonomic
characters for some symbiotic microorganisms. In particular, symbiosis has
been adopted as the sole basis for the classification of rhizobia (the bacteria
that can nodulate legumes). For example, in the 1984 edition of Bergey's
Manual of bacterial systematics the family Rhizobiaceae was erected to
accommodate all rhizobia.

Recent molecular studies have shown that the capacity to nodulate
legumes is not a sufficient character to classify rhizobia. Rhizobia comprise
at least two phylogenetically distinct groups. One is the genus *Rhizobium*,
related to *Agrobacterium*. *Rhizobium* appears to be a valid genus, in that
all *Rhizobium* species are more closely related to other *Rhizobium* species
than to members of other genera. The second group of rhizobia are the
bradyrhizobia, which are allied to *Rhodopseudomonas*. The genus *Bradyrhizobium* is not, however strictly valid because the bacteria capable of
nodulating legumes do not form a coherent group, distinct from nonsymbiotic taxa. Some strains of *Bradyrhizobium* are more closely related to the
type strain of *Rhodopseudomonas palustris* than to other *Bradyrhizobium*
strains (Young *et al.* 1991).

A further area of confusion is within the genus *Rhizobium*, because
species of this genus are defined by their host plant range. (It is not feasible
to classify bradyrhizobia in this way because most bradyrhizobia can infect
a wide diversity of legumes.) Early studies with crop legumes led to the
erection of a number of species (Table 5.1). However, not all strains of
Rhizobium conform to this scheme. Some *R. loti* strains can nodulate
lupins (members of the family Genistaceae, very distant from the Loteae);
and *Rhizobium* sp. strain Or191 can infect both *Medicago sativa* and
Phaseolus beans, even though it is not closely related to either *R. meliloti*
or *R. leguminosarum* bv. *phaseoli* (Eardly *et al.* 1992).

There are two reasons for the inadequacy of host plant range as a
taxonomic character in the classification of *Rhizobium*.

Table 5.1 Some species of *Rhizobium* and their host plants

Rhizobium	Plants
R. meliloti	Alfalfa (*Medicago*)
R. loti	Trefoils (*Lotus*)
R. leguminosarum	
bv. *viciae*	Pea (*Pisum*), vetches (*Vicia*)
bv. *trifolii*	Clovers (*Trifolium*)
bv. *phaseoli*	Beans (*Phaseolus*)

(1) Rhizobia are acquired from the soil by the root system of each legume plant. This provides regular opportunities for mixing of the legume and rhizobial lineages. As discussed in section 5.2.2, one would not expect host range to be a valid taxonomic character for such symbioses.

(2) The host plant range of *Rhizobium* is determined by genes borne on a plasmid, but the phylogenetic position of *Rhizobium* is determined by the chromosomal genes. The significance of this distinction depends on the extent of plasmid transfer between bacteria. If the plasmids were freely transmitted, so that they were randomly distributed among populations of *Rhizobium*, host range would have no taxonomic meaning whatsoever. If the plasmids were never mobilized, no phylogenetic distinction between the plasmid and chromosome would arise. The condition in many *Rhizobium* populations probably lies between these two extremes. Young and Wexler (1988) have shown that a natural field population of *R. leguminosarum* is divisible into several chromosomal lineages, each with an associated plasmid. Plasmid transfer between these lineages can, however, occur. Indistinguishable plasmids were occasionally identified with very different chromosomal backgrounds, and certain bacteria with different plasmids (and host plants) had similar chromosomal types.

5.3 HOW SPECIFIC ARE SYMBIOSES?

5.3.1 Specificity as a trade-off between specialization and the availability of partners

Specificity is moulded by the evolutionary history of the symbiosis. It can be viewed as the outcome of a trade-off, or compromise, between opposing selection pressures, one to broaden the range of acceptable partners and the other to become increasingly specialized.

The selective forces to specialize are inherent to biotic interactions (Futuyma and Moreno 1988). Let us consider a host species, which can form an association with three symbiont taxa, X, Y, and Z. Taxon X is a more effective symbiont that Y or Z, in that the host is more vigorous, fecund, etc. when interacting with X than with the alternative symbionts. The host would be under selection pressure: first, to choose X, in preference to Y or Z; and second to optimize its interactions with X (so enhancing the effectiveness of X), even if this entailed a further reduction in the effectiveness of Y and Z. In other words, an organism is under selection pressure to make the most of the best partner.

In vertically transmitted associations, there is no obvious opposing selection pressure to constrain the tendency to specialize (unless the relative effectiveness of X, Y, and Z varies with environmental conditions). However, for many associations involving symbionts acquired from the environment, the most effective symbiont (X) may not always be available. Then, provided the host performs better in association with Y or Z than without these symbionts, it is advantageous for the host to maintain a relatively broad specificity.

5.3.2 Variation in specificity: mycorrhizal fungi and plant roots

The specificity in some types of associations is variable, suggesting that the trade-off between the tendency to specialize and the hazard of remaining uninfected is far from uniform. As an example, the ectomycorrhizal fungi include several very specific species, such as *Alpova diplophloeus* on *Alnus* (alder), and *Suillus grevillei* on *Larix* (larch), but other fungal species have a broad plant range (e.g. *Amanita muscaria*, which associates with many conifers, and *Rhizopogon vinicolor*, which associates with both ericaceous shrubs and conifers).

In contrast to this variability in ectomycorrhizal fungi, all VA-mycorrhizal fungi are believed to have a very broad plant range. *Glomus* species, in particular, may be able to infect virtually any plant capable of forming a VA-mycorrhiza.

Allen (1991) has linked the specificity of mycorrhizal fungi to the characteristics of the plant community utilized by the fungal mycelium. *Alnus* is commonly the dominant tree in disturbed, wet habitats, such as river margins, and *Alpova diplophloeus* (an ectomycorrhizal fungus with narrow specificity) can persist, assured that compatible plant roots are available. In contrast, *Rhizopogon vinicolor* inhabits soils supporting mixed vegetation of ericaceous shrubs and conifers, and it is undoubtedly advantageous for the fungus to utilize both plant groups. The selection pressure against specialization (narrow specificity) may be even greater for VA-mycorrhizal fungi associated with relatively short-lived herbaceous plants. When a plant dies, the continued persistence of the fungal mycelium depends on its capacity to 'tap into' a nearby plant, which in many habitats is likely to be a different species.

5.3.3 Specificity in the laboratory and in nature: root nodules in legumes and algal symbionts in flatworms

The range of partners with which an organism forms an association can be assessed in two alternative ways: to challenge the organism with different potential partners under laboratory conditions, and score the combinations that generate a symbiosis; and to identify all the partners with which the organism associates in field conditions.

The specificity of a symbiosis commonly appears to be narrower in natural conditions than in the laboratory. Two factors that may contribute to this difference are considered here.

(a) *Only a limited range of potential partners are available to the organism in field conditions*

For some symbioses between crop legumes and rhizobia, this issue is of immense agricultural importance. Considerable research effort has been directed to the production of effective rhizobial strains for legume crops. The strains often fail to persist in natural soils, however, and the legume's choice of rhizobia is limited to the indigenous (and often relatively ineffective) populations in the soil. The inocula of effective rhizobia may be poor competitors with the indigenous rhizobia, either in relation to soil resources (e.g. carbon compounds) or for infection sites on the plant. Compounding this, they may be less able to thrive in natural soils than the indigenous rhizobia. Rhizobial strains are known to vary widely in their tolerance of low soil pH, water content, or extremes of temperature, and these abiotic factors may be as important as symbiosis-related factors in determining the range of rhizobia available to the legume (Alexander 1985).

(b) *An organism may be able to form a symbiosis with a variety of taxa, but will select one taxon when a choice is available*

Preference for particular partners occurs in *Convoluta roscoffensis*. *C. roscoffensis* is an acoel flatworm, which contains algal symbionts of the genus *Tetraselmis*. The life cycle of *C. roscoffensis* is shown in Fig. 5.5. Juvenile animals are alga-free, and they become infected when they feed

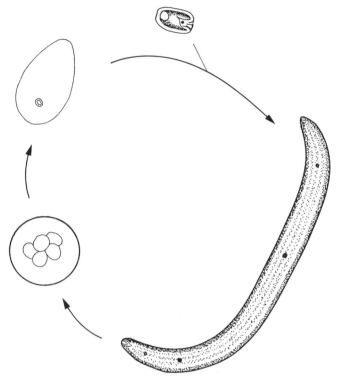

Fig. 5.5 The life cycle of the intertidal flatworm *Convoluta roscoffensis*. The adult animal contains green algae of the genus *Tetraselmis*. *C. roscoffensis* reproduces sexually, and the eggs are deposited into sand. The algae are not transmitted vertically, and juvenile animals hatching from the eggs are alga-free. The symbiosis is established when juveniles feed on *T. convolutae* in the sand.

on the appropriate algae. Laboratory cultures of juveniles ingest and form a symbiosis with any species of the green alga *Tetraselmis*. However, most natural populations of *C. roscoffensis* contain just one algal species, *T. convolutae*, even though a variety of *Tetraselmis* species are present in the immediate environs of *C. roscoffensis* populations.

A plausible explanation for the discrepancy between the specificity of *C. roscoffensis* for its algal symbionts in the field and laboratory comes from experiments of Provasoli *et al.* (1968) and Douglas (1983), showing that *C. roscoffensis* prefers *T. convolutae* to alternative *Tetraselmis* species. For example, *C. roscoffensis* can form an association with both *T. convolutae* and *T. marinus*, but when the animals are exposed to both species, either in sequence or simultaneously, only *T. convolutae* persists (Table 5.2). By mechanisms that are not understood, *T. marinus* is discriminated against and lost when both algae are present in the tissues of *C. roscoffensis*.

Table 5.2 Choice of algal symbionts by the flatworm *Convoluta roscoffensis*

Juvenile *C. roscoffensis* were incubated with algal cells (either *Tetraselmis convolutae* (TC) or *T. marinus* (TM)) at a density of 10^4 cells ml^{-1} for 2 days, and then washed free of exogenous algal cells. The symbiosis was allowed to develop, and the identity of algal cells in the animals was scored 30 days later. At least 10 animals were scored for each of the 5 treatments. Without exception, *T. convolutae* was selected as the symbiont when it was available. In cultures lacking this alga, the animals formed an association with the alternative alga, *T. marinus*.

Treatment of animals	Algal symbiont in animals at end of experiment
TC (2 days)	TC
TM (2 days)	TM
TC + TM (mixed culture, 2 days)	TC
TC (2 days), then TM (2 days)	TC
TM (2 days), then TC (2 days)	TC

(Unpublished data of A. E. Douglas).

Under laboratory conditions, *C. roscoffensis* grows faster and reaches sexual maturity at a younger age when infected with *T. convolutae* than with *T. marinus*. In this way, *C. roscoffensis* can select the most effective symbiont available.

5.3.4 The diversity of organisms in a single association

The ability of *C. roscoffensis* to select a single algal species, *T. convolutae*, from a mixture of potentially acceptable *Tetraselmis* species is not unique. The population of symbionts in a single host individual is remarkably uniform, and possibly clonal, for a variety of associations, including *Symbiodinium* in marine Cnidaria (Rowan and Powers 1991), luminescent bacteria in fish (Haygood *et al.* 1992), and chemosynthetic bacteria in bivalves (Eisen *et al.* 1992).

The genetic uniformity of symbionts in a single host individual may not only reflect the capacity of hosts to select the most effective symbiont. It may be advantageous to maintain a low diversity of symbionts, irrespective of their relative effectiveness. Different symbiont genotypes would be expected to have slightly different nutritional requirements, and mixed symbiont populations could therefore deplete different host resources independently. Genetically identical symbionts would have relatively uniform resources requirements, for which they may even have to compete. In other words, genetically homogenous symbionts may cost less to maintain than heterogenous symbiont populations.

The one situation in which this argument does not apply is where the symbiont (or other partner) derives nutrients from the external environment. It is then in the host's interest to maximize the range of environmental resources utilized. The two relevant associations are: first, the mycorrhizas in plants, where the fungi transfer mineral nutrients from the soil to the plant roots; and second, the gut symbioses in herbivorous animals, in which cellulolytic microorganisms utilize plant material ingested by the animal. Herbivorous mammals bear hundreds of microbial species (see Chapter 4, section 4), and the roots of a single plant may bear a variety of mycorrhizal fungi (see Fig. 1.3). For example, Allen (1991) scored 8 species of VA-mycorrhizal fungi associated with a single plant of *Ammophila breviligulata*; and in a study of the grass *Festuca*, Molina *et al.* (1978) identified two or more species of VA-fungus in over 80 per cent of plants examined. VA-fungal species vary in their hyphal characteristics, and multiple infections may enable plants to acquire soil resources under different environmental conditions.

5.4 RECOGNITION IN SYMBIOSIS

Let us return to the flatworm *Convoluta roscoffensis*, and the evidence in Table 5.2 that this host can discriminate between algae, selecting just one species to the exclusion of alternative taxa. The mechanisms underlying this process are obscure, but the animal is believed to recognize the appropriate algal cells, both at initial contact and in the developing symbiosis. Perhaps the surface of *Tetraselmis* species (and particularly *T. convolutae*) bears molecules that interact with complementary molecules on animal membranes. Definitive evidence for specific molecular interactions has not, however, been obtained for *C. roscoffensis* or most other associations.

The symbiosis between legumes and rhizobia (especially *Rhizobium*) is the only system in which host–symbiont interactions leading to the formation of a symbiosis are at least partly understood. Two compounds, produced in sequence, are involved: a plant flavonoid and a sugar derivative (called a nod factor) synthesized by the rhizobia. As illustrated in Fig. 5.6, the legume root releases a flavonoid, which interacts with a protein of the rhizobia, called NodD, causing other *nod* genes to be expressed. This leads to the synthesis of the nod factor, which triggers nodule formation by the legume.

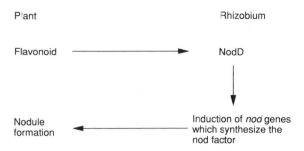

Plant   Rhizobium

Flavonoid ⟶ NodD

Nodule   Induction of *nod* genes
formation   which synthesize the
                          nod factor

Fig. 5.6 Molecular signalling between roots of leguminous plants and rhizobia in the soil.

The following section describes how this breakthrough in our understanding of the symbiosis has come from the genetic dissection of nodulation.

5.4.1 **The nodulation (*nod*) genes in rhizobia**

Two experimental approaches have been very important in studies of the genetic basis of recognition in rhizobia–legume symbioses: the use of rhizobia mutants and the use of reporter genes.

(a) *The study of rhizobia mutants with impaired capacity to nodulate legumes.*
This has led to the identification of *nod* genes, which control both the early stages in nodulation and host plant range of the rhizobia.

(b) *The use of reporter genes (e.g.* lacZ *of* E. coli *to investigate* nod *gene expression.*
The products of *nod* genes are either unknown or difficult to assay. If the *lacZ* sequence is inserted into the coding region of a *nod* gene, translation of the gene can be scored by the generation of the *lacZ* product, beta-galactosidase, for which a simple colorimetric assay is available.

The organization and expression of *nod* genes have been reviewed by Fisher and Long (1992) and Schlaman *et al.* (1992). In *Rhizobium*, the *nod* genes are clustered together on the Sym plasmid which also bears the genes for nitrogen fixation, *nif* and *fix*). The *nod* genes are grouped into several operons, and the promoter region of each operon includes a highly conserved sequence of about 50 kb, known as the *nod*-box. For example, most strains of *R. leguminosarum* bv. *viciae* have 13 nod genes in 5 operons, over a 10-kb stretch of the Sym plasmid (Fig. 5.7). *Nod* genes have also

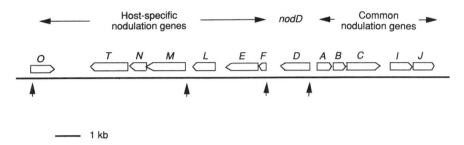

Fig. 5.7 Genetic organization of *nod* genes in *Rhizobium leguminosarum* bv. *viciae*. The genes are shown with points in the direction of transcription. The arrows indicate the positions of nod boxes. (Redrawn from Schlaman *et al.* (1992))

been identified in *Bradyrhizobium*, but in these rhizobia they are chromosomal and not tightly clustered.

NodD is of central importance to the interaction of rhizobia with legumes, and this gene is considered first.

5.4.2 *NodD* and the interaction between the NodD protein and plant flavonoids

NodD is the only *nod* gene expressed in freeliving rhizobia, and it controls the expression of all other *nod* genes. Its product (a 33-kd protein known as NodD) is a member of the LysR family of transcriptional-activating proteins, and it is active in the presence of certain flavonoids. A compatible flavonoid binds specifically to NodD, causing it to undergo a conformational change. As a result, NodD binds to the *nod*-box of the other *nod* operons, and this promotes transcription.

Different legumes release different flavonoids. For example, alfalfa and clover produce the flavone luteolin, while soybean produces the isoflavone daidzein, and peas produce the flavonone naningenin (Fig. 5.8(a)). There is also variation in the responsiveness of NodD proteins of the rhizobia. Daidzein is a potent inducer of *nod* genes in *B. japonicum* (the symbiont of soybean), but inhibits induction in *R. leguminosarum* bv. *trifolii* (the symbiont of clovers). The match between the plant flavonoid and rhizobial *nodD* is a vital component of the host plant specificity of rhizobia, because *nod* genes are expressed only when rhizobial cells are close to roots producing a flavonoid compatible with *nodD*.

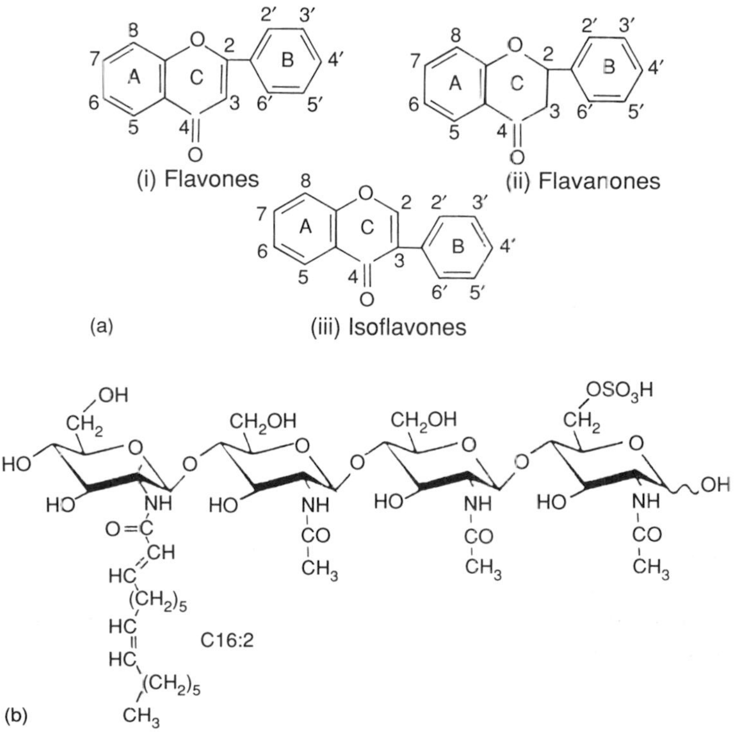

Fig. 5.8 Molecules mediating legume–rhizobium recognition. (a) Flavonoids produced by the legume root. (i) Flavone ring structure. Flavones include: luteolin (5, 7, 3′, 4′-tetrahydoxylflavone) and apigenin (5, 7, 4′-trihydroxyflavone). (ii) Flavonone ring structure. Hesperitin (5, 7, 3′-trihydroxy, 4′-methoxyflavanone) is a flavonone. (iii) Isoflavanone ring structure. Daidzein (7, 4′-hydroxyisoflavone) is an isoflavanone. (b) The nod factor NodRm-IV(S) produced by rhizobia, in response to the flavone luteolin.

5.4.3 *nod* genes and the synthesis of nod factors

A nod factor is a compound released from rhizobial cells that causes the root hairs of a host legume to curl and induces the nodule meristem (see Fig. 5.6).

The nod factor is an oligosaccharide, specifically a beta-1,4-oligomer of N-acetylglucosamine with an acyl chain at the reducing end (Fig. 5.8(b)). The detailed structure of the nod factor varies between rhizobial species, and some rhizobia may synthesize several slightly different nod factors.

A simple bioassay was crucial to the identification of the nod factors. It was found that filtrates from rhizobial cells (induced by the appropriate flavonoid) cause root hair curling only of plants which the rhizobium could nodulate. Leronge *et al.* (1990) showed that a single compound in the filtrate of induced *R. meliloti* was responsible for root-hair curling, and they purified this compound, which is now known as the nod factor NodRm-IV(S) (see Fig. 5.8(b)).

Rhizobia with mutations in *nod* genes (especially *nodA* or *nodC*) neither produce the nod factor, nor induce root-hair deformation in the host legume. These experiments indicate that the nod factor is synthesized by the products of *nod* genes.

Much recent research on nodulation by rhizobia has concerned the contribution of individual *nod* genes to the production of the nod factor. *NodABC* genes code for proteins that are essential for the nod factor and nodulation, and both these genes and *nodIJ* are interchangeable between different *Rhizobium* species. For example, if *nodA* from *R. leguminosarum* is introduced to a mutant of *R. meliloti* lacking a functional *nodA*, the mutant recovers its capacity to nodulate its host plant, *Medicago* (but not the host of the *R. leguminosarum* strain). The functions of the products of *nodAB* are unknown. *NodC* may code for an enzyme that polymerizes the constituent sugars of the nod factor.

The other *nod* genes are known as host-specific nodulation genes because they are not interchangeable between legume hosts. Several host-specific genes are present in all *Rhizobium* species (e.g. *nodFE*, *nodL*, *nodM*) but others are particular to certain species or biovars (e.g. *nodT* in *R. leguminosarum* bv. *viciae* and *nodHPQ* in *R. meliloti*). These host-specific *nod* genes control the specific features of the nod factor structure, e.g. the length of the acyl chain and whether the sugar residues bear acetyl or sulphate residues; and it is these features that determine whether a legume responds to the nod factor. For example, *NodFE* are responsible for the synthesis of the acyl chain (they code for an acyl carrier protein and fatty acyl synthetase, respectively); and *nodHPQ* probably add sulphate to the C-6 atom of the terminal sugar of the nod factor, NodRm-IV(S) in *R. meliloti*.

There remains the nature of the interaction between the nod factor and the plant, and how this interaction leads to nodule organogenesis. Fisher and Long (1992) propose that legumes may bear a family of receptors for nod factors. The plant response to the binding of a nod factor to a putative receptor is obscure, but it may involve modification of the levels of plant auxins or cytokinines (these are plant hormones).

5.4.4 Flavonoids and oligosaccharides in plant-microbial interactions

To summarize, the first interaction between compatible rhizobia and legumes involves a flavonoid–oligosaccharide signal sequence. This sequence shapes the specificity of the symbiosis because the rhizobium NodD can respond to only a few flavonoids, and the plant can respond to only a restricted range of nod factors.

Plant flavonoids and microbial oligosaccharides are also important signals in plant–pathogen interactions. Microbial 'elicitor' compounds, which trigger plant defensive responses are sugars, often oligomers of N-acetylglucosamine, and the plant response to elicitor molecules involves the production of phytoalexins, including flavonoids. The recognition system in rhizobium–legume symbioses has probably evolved from antagonistic signal exchange between legumes and pathogenic microorganisms. It is possible that the ancestors of rhizobia were among these pathogens.

6

Regulation of microbial symbionts

In symbiosis research, the term 'regulation' has a very specific meaning. It refers to the mechanisms by which the relative biomass of the partners in an association are controlled. Under most conditions, symbioses are stable. The symbionts occupy a predictable proportion of the volume or biomass of the association, and they are restricted to particular regions of the host body. For example, the symbiotic *Chlorella* in hydra are harboured in hydra digestive cells (see Fig. 1.2(b)), and they occupy 10 per cent of the digestive cell volume; and the bacteria in aphids account for 60 per cent of the volume of the insect's mycetocytes, and do not colonize any other aphid cells.

The relative biomass of the organisms in symbioses is stable because the symbionts proliferate in parallel with growth of the host. However, this apparent harmony masks a conflict between the partners. The intrinsic growth rates of microorganisms are much higher than those of multicellular eukaryotes (particularly plants and animals). For example the doubling time of *Chlorella* in culture is 10 hours but their hydra hosts double every 8–10 days. Similarly, the bacteria in aphids are allied to enteric bacteria with a capacity to divide every 20 minutes, while aphids double in biomass no faster than once per 3–4 days. Regulation concerns the ways in which this conflict between host and symbionts is resolved.

6.1 HOW SYMBIONT POPULATIONS ARE REGULATED

In the most associations, the symbionts are controlled by suppression of their growth and division, so that they and their host increase at the same rates. Symbionts can also be eliminated, either by expulsion from the host or by lysis. The contribution of these modes of regulation varies between associations, and with environmental conditions and developmental age of the host, as the following examples illustrate.

6.1.1 Symbioses between marine Cnidaria and the alga *Symbiodinium*

The *Symbiodinium* populations in corals, sea anemones, and other marine Cnidaria increase very slowly, with growth rates rarely exceeding 0.1 day^{-1} (equivalent to a doubling time of 10 days). Even so, the growth rate of their hosts is even lower. For example, the coral *Stylophora pistillata* in its natural habitat coastal waters of the Red Sea) increases by 0.001–0.005 d^{-1}, up to 100 times slower than its algal symbionts (Hoegh-Guldberg *et al.* 1987).

The relative biomass of *Symbiodinium* and their cnidarian host remains constant only because the excess symbiont cells are expelled or lysed by the host. However, the relative importance of expulsion and lysis in the regulation of *Symbiodinium* is uncertain, and may vary between cnidarian species and with environmental conditions.

Symbiotic Cnidaria are often observed to expel pellets of *Symbiodinium* via the mouth, but often the expelled symbionts represent a very small proportion of the algal population. For example, the algal cells expelled from *Stylophora* account for less than 4 per cent of the daily increase in the algal population (Hoegh-Guldberg *et al.* 1987).

Lysis of algal symbionts is not often observed. Most associations are, however, studied during the daytime, and this may give misleading results. Fitt and Cook (1990) have found that in a tropical hydroid, *Myrmionema ambionesne*, lysis of *Symbiodinium* cells is appreciable, and occurs mostly at night (Fig. 6.1(a)).

6.1.2 The mycetocyte symbiosis in aphids

The bacteria in aphids are transmitted vertically, and they colonize the mycetocytes as these cells differentiate in the aphid embryo. By the time it is born, an aphid contains 50–100 mycetocytes (varying with environmental conditions and species). These cells rarely divide in the larval or adult insect, but they do increase in size, especially in the young larvae (Douglas and Dixon 1987).

In young pea aphids, the mycetocytes are growing rapidly and approximately 15 per cent of the bacteria in the mycetocytes are in a dividing condition (Whitehead and Douglas 1993). (Unfortunately, the proliferation rate of the bacteria cannot be calculated from this value because it is not known how long the bacteria take to divide or to complete a full cell cycle.) As the aphid ages, both mycetocyte growth rate and the rate of bacterial division decline, so that when the aphid is 8 days old the percentage of dividing bacteria halves to approximately 7 per cent (Fig. 6.2), and this

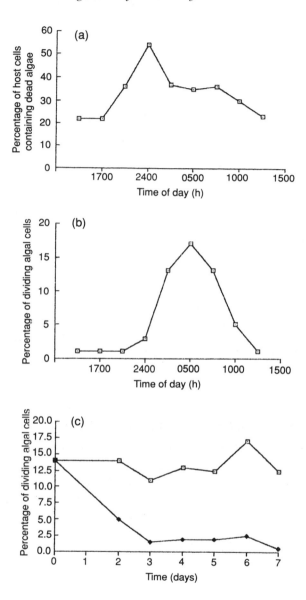

Fig. 6.1 Regulation of the *Symbiodinium* population in the marine hydroid *Myrionema ambionesne*. Natural populations of *M. ambionesne* display a diurnal rhythm with (a) peak of lysis at midnight (3 h after nightfall at 2100 h); (b) a peak of division at daybreak (0500 h). In the laboratory (c), the division rates of *Symbiodinium* depend on whether the hydroid host is fed (□) or starved (●). (Redrawn from Fitt and Cook (1990))

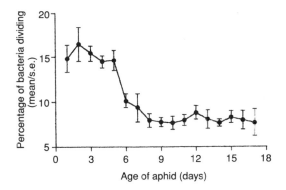

Fig. 6.2 Percentage of dividing bacteria in mycetocytes of the pea aphid *Acyrtho-siphon pisum*. (From Whitehead and Douglas (1993))

lower value is maintained to adulthood. Degenerating bacteria are rarely observed in larval aphids, and it is very likely that the bacteria are controlled primarily through their growth and division rates.

The aphid can, however, destroy its symbionts. In adult aphids, the mycetocytes break down. The bacteria do not escape from the degenerating mycetocytes, but are lysed with the host cell (Douglas and Dixon 1987).

6.1.3 The association between legumes and rhizobia

Proliferation of rhizobia in legume nodules is sharply curtailed, and most *Rhizobium* cells in infected cells of the nodule appear to have lost their capacity for division. These non-dividing *Rhizobium* cells increase in size and become irregularly shaped. *Bradyrhizobium* retain a greater capacity for division than *Rhizobium*, and they change little in shape or size. Lysis of rhizobia in the nodule is rarely observed until the nodule tissue starts to senesce, when many of the intracellular rhizobia may be destroyed (Brewin 1992).

6.1.4 The lichen symbiosis

Lichens grow exceptionally slowly, and most of the increase is restricted to the apex or margin of the thallus. In the foliose and fruticose lichens, very few symbiont cells are lysed, and the proliferation of symbionts is linked to their position in the thallus. For example, in the foliose lichen *Parmelia saxatilis*, 15–20 per cent of the algal cells are dividing at the growing margin, but only 1 per cent of the cells are dividing at the thallus centre (Greenhalgh and Anglesea 1979).

6.2 THE MECHANISMS OF REGULATION

6.2.1 Control of symbiont proliferation by nutrient-limitation

Hosts are believed to control the rate of symbiont proliferation by controlling the supply of nutrients to the symbionts. In other words, the growth and division of symbionts is nutrient-limited.

The most persuasive evidence for this hypothesis comes from studies of alga–Cnidaria symbioses. In these associations, growth and division of the algal cells usually occurs soon after the host feeds. The hydroid *Myrionema ambionesne* studied by Fitt and Cook (1990) feed by night, and the population of *Symbiodinium* divides early in the following morning (see Fig. 6.1(b)). If the hydroid does not feed, the algae do not divide (Fig. 6.1(c)). Similarly, symbiotic *Chlorella* in hydra divide only when their host is fed. The identity of the limiting nutrient is uncertain, but there is evidence that the algal cells may be deficient in both nitrogen and carbon.

(a) *The evidence that symbiont proliferation is limited by host-derived nitrogen*

Symbiotic algae have characteristics of nitrogen-deficient cells. For example, *Chlorella* in hydra have a low nitrogen content and accumulate nitrogenous compounds at very high rates (Table 6.1(a)). McAuley (1987) has suggested that hydra regulate the proliferation of *Chlorella* by controlling the supply of nitrogen, possibly as amino acids, across the symbiosome membrane.

Elevated concentrations of ammonia or other inorganic nutrients in the medium can cause an increase in the algal population. When sea water is supplemented with ammonia, the population of *Symbiodinium* in corals increases (Table 6.1(b)), causing the coral tissues to become chocolate-brown, the colour of *Symbiodinium* cells. One interpretation of these results is that the host's control over nutrient supply to the symbionts is overwhelmed by the high levels of exogenous nutrients. However, the high concentrations of ammonia may also have a general deleterious effect on cnidarian tissues, and the high symbiont populations may reflect a general malaise in the host.

(b) *The evidence that symbiont proliferation is carbon-limited*

Symbiotic algae release most of their photosynthetically fixed carbon to the host (Chapter 4, section 1.1). As a result, they retain very little carbon,

Table 6.1 Nutritional status of symbiotic algae

(a) The nitrogen content and amino acid uptake rates of symbiotic *Chlorella*.

Chlorella	Mean nitrogen content of algae (fg μm^{-3} cell volume)	Mean rate of alanine uptake (fmol cell^{-1} h^{-1})
In symbiosis	0.44	1.8
N-replete cells in culture	0.59	0.5

(Data from McAuley (1987)).

(b) The effect of ammonia enrichment of seawater on algal density in corals.

The corals were incubated in seawater supplemented with 10–40 μM ammonia for 19 days, with corals in unenriched seawater as control.

Enrichment of sea water	Density of algae in corals (10^6 × no. cells mg^{-1} protein) mean ± s.e., 10 reps.	
	Stylophora pistillata	*Seriatopora hystrix*
None	0.55 ± 0.124	2.11 ± 1.025
Ammonia	1.49 ± 0.248	2.78 ± 1.55

Data from Hoegh-Guldberg and Smith (1987).

sufficient to fuel only low growth rates. This raises the possibility that photosynthate release may contribute to the control of algal proliferation. The principal evidence comes from green hydra infected with *Chlorella* strains varying in their capacity to release photosynthate. One *Chlorella*, known as strain NC64A, releases very little photosynthate to the host, and this strain has considerably higher division rates than strains (such as 3N8/13–1) with high rates of photosynthate release (Table 6.2).

(c) *Conclusions*

The suggestions that algae in Cnidaria are nitrogen-deficient (Table 6.1) and carbon-deficient (Table 6.2) are not necessarily alternatives. They may be closely linked characteristics of symbiotic algae, and a consequence of the mechanism by which cnidarian hosts trigger photosynthate release in the algal cells. According to the metabolic imbalance hypothesis (Chapter 4, section 1.3), the algae should release photosynthate because they are deficient in nitrogen.

Table 6.2 The relationship between photosynthate release and division rates of symbiotic *Chlorella* in hydra

Chlorella strain[1] in hydra	Percentage of photosynthate released to host	Percentage of dividing alga cells
3N8/13–1	55–60	3.1
NC64A	15–20	6.0

[1] The two *Chlorella* strains were obtained from the ciliate protist *Paramecium bursaria* and introduced, experimentally, to the hydra. Their photosynthetic rates (per unit chlorophyll) in symbiosis with hydra were closely similar.

(Data: A. E. Douglas and P. J. McAuley.)

6.2.2 **The link between symbiont proliferation and host cell division**

Two issues arise when a host cell divides.

Is the division of symbionts synchronized with host cell division?
How is the segregation of symbionts to the host daughter cells controlled?

An important factor influencing the precision of regulatory mechanisms at host-cell division may be the number of symbionts per host cell. Let us consider a host cell containing just one symbiont and about to divide. If the host daughter cells are each to obtain a symbiont, symbiont division must be triggered by the onset of host cell division, and one of the symbiont daughter cells must be partitioned to each host daughter cell. If the single symbiont fails to divide, one host daughter cell would be symbiont-free; and if segregation of the recently divided symbionts were random, the probability that both symbionts were allocated to one of the two host cells would be 0.5. As the number of symbionts per host cell increases, so the 'risk' of generating a symbiont-free host daughter cell by random partition of symbionts declines, from 0.5 for two symbionts per cell to 0.002 for 10 symbionts and to 2×10^{-6} for 20 symbionts (Fig. 6.3).

Only very limited data are available, but they support the notion that host cells with low symbiont numbers have precise regulatory mechanisms at division. *Paulinella* is a protist that divides by binary fission and has a single intracellular cyanobacterial symbiont. Host–symbiont division is synchronous, and the cyanobacteria are segregated, one to each daughter cell of *Paulinella*. The gastrodermal cells of hydra contain (on average) 20 *Chlorella* cells. The *Chlorella* divide only at hydra cell division, but their segregation to hydra daughter cells is random (McAuley 1990). Each mycetocyte in cockroaches contains thousands of bacteria. In larval

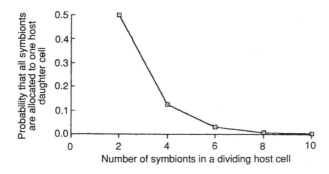

Fig. 6.3 Segregation of intracellular symbionts at host cell division. If the symbionts are partitioned at random, the probability that all symbionts are allocated to one host daughter cell declines rapidly with increasing number of symbionts per cell.

cockroaches, the mycetocytes divide only when the insect moults. The intracellular bacteria divide throughout larval development, and they are partitioned randomly between the daughter mycetocytes (Brooks and Richards 1955).

In a few symbioses, the division and segregation of symbionts is linked to host organelles and not the host cell as a whole. The anaerobic ciliate *Plagiopyla frontata* contains several thousand methanogenic bacteria, and each bacterium is sandwiched between two hydrogenosomes (Fig. 6.4). Fenchel and Finlay (1991b) have shown that the regular alternation between hydrogenosome and methanogenic symbiont is maintained by synchronous division of adjacent organelles and symbionts and by controlled partition of the bacteria between hydrogenosomes. The underlying mechanisms are obscure, but presumably involve the ciliate's cytoskeletal system.

6.2.3 **Regulation by destruction of the symbionts**

The cells or tissues of a different organism are a dangerous place to live. If a symbiotic microorganism is to persist, it must avoid or suppress the nonspecific defences of its host against foreign organisms. For intracellular symbionts, the symbiosome membrane is a special hazard, because this membrane is potentially able to fuse with lysosomes, containing hydrolases and other degradative enzymes. A few intracellular symbionts, including *Chlorella* in hydra (Hohman *et al.* 1982), have been demonstrated to avoid lysosomal fusion, but none is known to survive lysosomal attack.

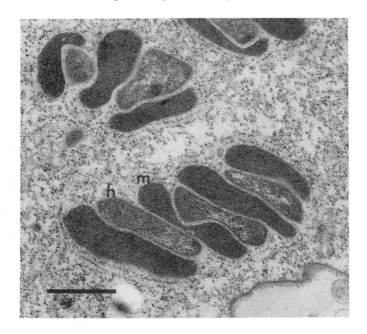

Fig. 6.4 Association between methanogenic symbionts and hydrogenosomes in the ciliate *Plagiopyla frontata*. m, methanogen; h, hydrogenosome. Scale: 2.75 μm. From Fenchel and Finlay (1991*b*))

From the perspective of the host, very precise control over its own destructive capabilities is essential. Particularly for hosts which derive nutrients from living cells of the symbiont (i.e. biotrophically), symbiont lysis is akin to 'killing the goose that lays the golden eggs'. A lysed symbiont, like a dead goose, may be a source of nutrients, but the long-term biotrophic supply of nutrients cannot be recovered. A host may lyse small numbers of symbionts, to cull a rapidly increasing population (as may occur in Cnidaria; see section 1.1), or selectively to destroy individual symbiont cells that fail to supply nutrients. Widespread destruction of symbionts may accompany programmed breakdown of the association, as in the aphid mycetocytes and legume nodules.

There is no direct information on the mechanisms by which symbiosome–lysosome fusion and symbiont lysis are controlled.

6.3 THE CARRYING CAPACITY OF HOSTS

Let us return to the green hydra and the various *Chlorella* that release different amounts of photosynthate. As described in section 2.1 and

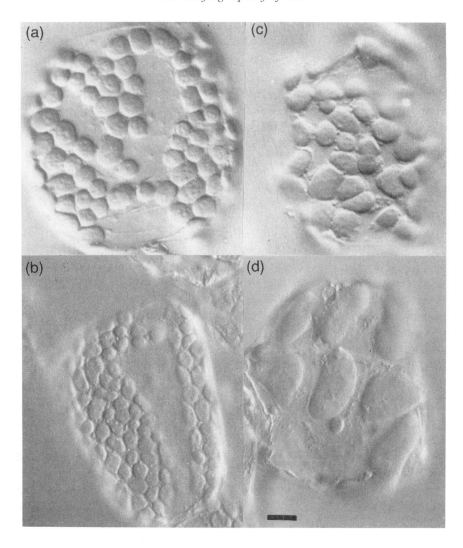

Fig. 6.5 Relationship between division and size of plastids in *Arabidopsis thalliana* Light micrographs of fully expanded leaf mesophyll cells (scale: 10 μm), showing plastids of (a) wild-type; (b) mutant *arc1* with plastids that divide very rapidly; (c) mutant *arc2* with slowly dividing plastids; (d) mutant *arc3* with plastids that are virtually incapable of division. (Reproduced from Pyke and Leech (1992))

Table 6.2, *Chlorella* strains (such as NC64A) with very low rates of photosynthate release proliferate rapidly. They do not, however, overgrow the host, but occupy a similar proportion of the hydra gastrodermal cell volume as *Chlorella* strains that release more than half their photosynthetic carbon

(Douglas and Smith 1984), the stable alga:host volume ratio is maintained because the excess cells of strain NC64A are expelled from the hydra.

These observations indicate that the relative biomass of host and symbionts is not a simple consequence of the relative capacity of host and symbionts to increase, but is fixed independently of the symbionts' proliferation rate. The biomass of symbionts supported in an association can be considered to reflect the carrying capacity of the host.

The factors that determine the carrying capacity of a host are obscure. The one certain issue is that the carrying capacity is set according to the biomass or space occupied by the symbionts, and not by their number.

The most striking evidence comes from studies, not with intracellular symbioses, but on the relationship between plant cells and their plastids. Pyke and Leech (1992) have investigated the proliferation of plastids in leaf mesophyll cells of *Arabidopsis*, as the plant cells expand during leaf development. They have isolated three nuclear mutants with aberrant rates of chloroplast division: in *arc1*, the chloroplasts divide very rapidly; in *arc2*, they divide slowly; and in *arc3*, chloroplast division is virtually absent. These mutations do not affect the proportion of the mesophyll cell volume allocated to the chloroplasts, because the variation in chloroplast proliferation is compensated for, by variation in size of the chloroplast. The mesophyll cells in *arc1* contain many small chloroplasts; *arc2* has a smaller number of large chloroplasts; and the few chloroplasts in *arc3* are (for organelles) enormous (Fig. 6.5).

7
The ecological impact of symbiosis

Symbiosis is rarely considered in the ecological literature. There are two reasons (neither fully justified) for the neglect of this topic by ecologists.

(a) Symbioses are often assumed to be mutualistic. When mutualistic interactions are modelled, the cooperating organisms proliferate uncontrollably, never reaching equilibrium (see review by Boucher 1985). This has led to the belief among some ecologists that symbioses are fundamentally unstable and prone to break down when exposed to environmental change or stress. In some ecological texts, symbioses are claimed to be restricted to supposedly stable environments in the tropics.

This line of reasoning is fallacious. As considered in Chapter 1, organisms do not always benefit from symbioses, and the associations are unlikely to be perfectly mutualistic. The view that symbioses are unstable is also belied by the facts. Some symbioses occur in environmentally variable habitats, including high latitudes (e.g. lichens, ericoid mycorrhizas); and many have persisted for millions of years. As examples, some groups of lichens were almost certainly established within the Permian period (>250 million years ago) (Hawksworth and Hill 1984); the VA-mycorrhizas may have arisen in the early Devonian (c. 400 million years ago) (see Chapter 2, section 1.3 and Pirozynski and Malloch 1975); and mitochondria, present in most eukaryotic cells, are believed to have evolved from symbiotic bacteria more than 10^9 years ago (see Chapter 8).

(b) It is widely perceived that the most appropriate methods for studying symbioses are physiological or biochemical, and not ecological. This viewpoint is only partly valid.

Ecological principles have, to date, contributed little to our understanding of interactions between organisms in symbiosis. One ecologically based approach is to treat hosts as examples of extreme and stressful habitats. This provides an interesting perspective, but has not led to substantive new insights into host–symbiont interactions.

Ecological approaches are, however, vital to establish how organisms in a symbiosis interact with other organisms and the physical environment.

In the first part of this chapter, the significance of symbioses in nutrient cycling (especially in the carbon, nitrogen, and phosphorus cycles) is reviewed. A variety of associations is considered, including coral reefs, lichens, and the nitrogen-fixing and mycorrhizal associations in plants.

The latter part of the chapter concerns plant–root symbioses with mycorrhizal fungi and nitrogen-fixing bacteria, and their impact on ecological processes in natural vegetation. I have chosen these systems because the symbioses are not universal in plants. For organisms that are invariably in symbiosis, it is very difficult to dissect out the significance of the symbiosis. As an example, the coral–alga symbiosis is fundamental to coral reefs, but the ecological impact of the symbiosis is most appropriately described in terms of the functioning of the reef ecosystem. Similarly, herbivorous mammals are dependent on their association with cellulose-degrading microorganisms, but the ecological impact of the association is equivalent to the impact of herbivory. Both coral reefs and herbivory are amply considered in many ecological texts, and do not warrant repetition here.

7.1 SYMBIOSIS AND THE FLUX OF ENERGY AND NUTRIENTS THROUGH ECOSYSTEMS

7.1.1 The fate of primary production

Symbioses are of particular importance in the utilization of photosynthetic carbon and the relationship between primary producers (plants, algae, and cyanobacteria) and their consumers.

Let us consider algae and cyanobacteria first (Fig. 7.1(a)). A variety of aquatic protists and animals, and the lichenized fungi on land, form symbioses with algae and cyanobacteria (see Tables 2.1 and 2.2). Up to 90 per cent of the photosynthate (i.e. net primary production) of the symbionts is transferred biotrophically from living cells of the photosynthetic partner to the heterotrophic hosts.

Some of the associations are of considerable ecological importance. For example, lichens are the most important primary producers in habitats covering 8 per cent of the land surface (1.2×10^7 km^2); these include Antarctica, many arctic–alpine habitats, and some deserts (e.g. the Namib of SW Africa) (Kappen 1988). A second example is the coral–alga symbiosis, which underpins all coral reef ecosystems, accounting for 6×10^5 km^2 of the world's oceans. The primary productivity of coral reefs is 1000–5000 g C m^{-2} year^{-1}, two orders of magnitude greater than of open tropical waters (20–50 g C m^{-2} year^{-1}). On a global basis, the algae in corals fix 3–4 times more carbon than the phytoplankton in all the nutrient upwelling regions of the open ocean (Hatcher 1990).

Terrestrial plants are also utilized biotrophically through symbiosis (Fig. 7.1(b)). Mycorrhizal fungi obtained 10–20 per cent of the photosynthate of plants. This represents a major sink for terrestrial primary production, simply because most plants are mycorrhizal (Brundrett 1991). More than 90 per cent of plant species in tropical forests (both rainforests and season-

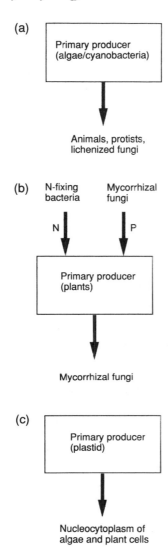

Fig. 7.1 Symbiosis and the flux of carbon, nitrogen, and phosphorus. The primary production of (a) algae and cyanobacteria, (b) terrestrial plants, and (c) plastids, is utilized widely by heterotrophs through biotrophic symbiosis (arrows). Some plants also obtain nitrogen and phosphorous through biotrophic symbiosis.

ally dry forests) have mycorrhizal fungi, and mycorrhizal vegetation is the norm in temperate and boreal forests, savannah, temperate grasslands, and Mediterranean-type habitats. The net primary production of vegetation in these habitats is 10^{14} kg C year^{-1} (calculated from data in Whittaker 1975), giving an annual global transfer of $1-2 \times 10^{13}$ kg C from plants to mycorrhizal fungi.

It is also appropriate to examine the nature of primary producers. The plastids in algae and plants are almost certainly derived from symbiotic cyanobacteria, and the nucleocytoplasm of these eukaryotes can therefore be considered to obtain photosynthate through a biotrophic symbiosis, in the same way as in symbioses of algae or cyanobacteria with lichenized fungi or corals (compare Fig. 7.1(a) and 7.1(c)). From this standpoint, virtually all primary production on land is symbiotic. (Nonsymbiotic systems (i.e. cyanobacteria) are, however, important primary producers in open waters of the ocean.)

Primary production is utilized by two routes in addition to biotrophic symbioses: detritivores and herbivores. In many terrestrial ecosystems, detritivores consume *c.* 70 per cent, and herbivores 10 per cent of net primary production. Symbiosis plays little part in these processes. It is not implicated in detritivory; and only mammalian herbivores (and not most insect herbivores) utilize cellulose-degrading symbionts to obtain carbon from ingested plant material (see also Chapter 4, sections 4 and 5).

7.1.2 Symbioses and the nitrogen and phosphorus cycles

Nitrogen-fixing bacteria (particularly rhizobia and *Frankia*) and mycorrhizal fungi associated with plants play a crucial role in the nitrogen and phosphorus cycles in terrestrial ecosystems.

Globally, an estimated 140×10^9 kg N year^{-1} is fixed by bacteria on land (Rosswall 1983), and virtually all of this terrestrial fixation is mediated by symbionts of plants. For certain soils, nitrogen-fixing plant symbioses represent the single most important source of nitrogen, as the following examples illustrate.

(a) Temperate agricultural grassland often contains nodulated legumes, such as *Medicago* (lucerne) and *Trifolium* (clovers). With *Trifolium repens* at 10 per cent cover, a grassland sward may fix up to 50 kg N ha^{-1} year^{-1}, which is sufficient to support the nitrogen requirements of a self-sustaining grassland system (Sheehy 1989).
(b) Stands of actinorhizal trees (i.e. with *Frankia* as symbiont), such as *Alnus* along river banks or *Casuarina* on coastal sand dunes, fix nitrogen at rates between 50 and 120 kg ha^{-1} year^{-1}, accounting for more than 75 per cent of the total nitrogen input for these soils (Sprent and Sprent 1990).

In aquatic systems, most biological nitrogen fixation is nonsymbiotic (i.e. mediated by free-living bacteria). However, the nitrogen-fixing symbioses in the marine diatom *Rhizosolenia* and in the water-fern *Azolla* can be locally important. *Rhizosolenia* contributes up to 15 per cent of the total nitrogen fixed in many tropical and subtropical regions of the Pacific Ocean (Villareal 1987), and *Azolla* in rice fields of south-east Asia may fix 50–150 kg N ha^{-1} over 1–4-month growing seasons (Lumpkin and Plucknett 1980).

Phosphate and ammonium, both relatively immobile ions in soils, are accumulated by mycorrhizal fungi: phosphate by both VA- and ecto-mycorrhizal fungi (especially at low latitudes) and ammonium by ecto-mycorrhizal fungi in boreal forests. Much of the phosphorus and nitrogen is channelled biotrophically, directly to plants (Fig. 7.1(b)), and this contributes to the tight cycling of N and P, especially in forest and grassland communities (Wood *et al.* 1984; Sprent 1986).

Although it is accepted that mycorrhizal fungi have a major influence on the rate and pattern of nitrogen and phosphorus cycles in soil systems, many of the ramifications of their activities are not fully understood. For example, one possible consequence of the efficient transfer of N and P via mycorrhizas to plants is a reduced availability of these nutrients to saprophytic microbiota in the soil, resulting in retarded decomposition of litter (Gadgil and Gadgil 1971). Consistent with this possibility, saprophytic fungi are often relatively scarce in the immediate environment of mycorrhizal plants (Brundrett 1991), but other studies (e.g. Harmer and Alexander 1985) indicate that the exclusion of mycorrhizas has no influence on decomposition rates by saprophytic fungi in forest-floor soils.

In the following section, we shall turn to the direct effect of mycorrhizal fungi on vegetation.

7.2 MYCORRHIZAS AND THE STRUCTURE OF PLANT COMMUNITIES

Mycorrhizal fungi would be expected to influence the species composition of plant communities. The reasoning is as follows: (a) The availability of soil nutrients is an important factor shaping the structure of natural vegetation. (b) Mycorrhizas enhance mineral nutrient uptake by most plants, but there is considerable variation in the 'responsiveness' of different plant species to mycorrhizal infection. (As an explanation of responsiveness, let us consider two plant species, *A* and *B*. If the vigour of species *A* is promoted by a mycorrhizal fungus to a greater extent than species *B*, then *A* is said to be more responsive than *B* to the association.) (c) Consequently, mycorrhizal fungi would influence the outcome of competitive interactions between plants with different responsiveness. The competitive position of species *A* versus species *B* would be enhanced in soil bearing appropriate mycorrhizal fungi.

7.2.1 Mycorrhizas and competitive interactions between plants

An experiment of Fitter (1977) conducted on two grasses, *Holcus lanatus* and *Lolium perenne*, illustrates how mycorrhizal fungi can influence the outcome of competition between plants. Single seedlings of each species were raised together on low-phosphorus soils for 3 months. Both in soil that lacked VA-mycorrhizal fungus and in pots with a divider separating the two root systems, the two species were competitively balanced, with *Holcus* accounting for 52–58 per cent of the plant dry weight (Table 7.1). When the two species were allowed to share a common mycorrhizal mycelium (i.e. there was no divider), *Holcus* became competitively superior, representing 75 per cent of the plant biomass.

Further analysis indicated that *Holcus* derived phosphate more effectively from the mycorrhizal fungus than did *Lolium*, with the result that, in the presence of the fungus, *Holcus* had the competitive edge over *Lolium*.

7.2.2 Mycorrhizas and the species composition of vegetation

As an approach to investigate whether mycorrhizal fungi may influence the diversity of plant species, Grime *et al.* (1987) set up miniature swards

Table 7.1 Mycorrhizal fungi and competition between plants

Seedlings of two grasses *Lolium perenne* and *Holcus lanatus* (one of each species) were raised in low phosphorus soil, either containing or lacking VA-mycorrhizal fungus. The root systems of the two plants were either separated by a transparent plastic divider or allowed to intermingle and compete (no divider). The yield of each plant (dry weight) was quantified after 3 months, and the data are expressed as the proportion of total plant biomass represented by *H. lanatus* (i.e. values ⩾0.50 indicate that *H. lanatus* is competitively superior to *L. perenne*).

	Yield of *H. lanatus* as proportion of total	
	With mycorrhizal-fungus	No mycorrhizal-fungus
With divider	0.58	0.52
No divider	0.75	0.54

(termed *microcosms*) of grasses and forbs in soil that either contained or lacked VA-mycorrhizal fungi. In the microcosms lacking mycorrhizal fungi, the forbs were competitively subordinate and the grass *Festuca ovina* was dominant. In soils inoculated with the mycorrhizal fungi, the mycorrhizal forbs flourished, coming to represent a higher proportion of the total plant biomass than the *Festuca*. The plant community generated with mycorrhizal fungi was more diverse than in microcosms free of VA-mycorrhizal fungi.

The results from the microcosm experiments have been interpreted in terms of nutrient transfer between plants connected by a common mycelium of mycorrhizal fungus (see Fig. 1.3) (Grime *et al.* 1987). It is argued that a plant deficient in minerals or photosynthate would act as a sink, deriving phosphate from the fungus and photosynthate from fungal-connected plants. In this way, the mycorrhizal fungus would reduce the nutritional differences (and therefore competitive differences) between plant species, and promote coexistence.

It is, however, very uncertain that the flux of nutrients between plants via mycorrhizal fungi is sufficient to alter the competitive balance in this way. An alternative explanation for the increased species diversity on the mycorrhizal swards is that many of the forb species are more responsive to mycorrhizal fungi than the grass *Festuca*. In other words, the mycorrhizal fungi enhanced the capacity of some forbs to acquire phosphate to a greater extent than for *Festuca*, and so the competitive advantage of *Festuca* over the forbs was diminished in the mycorrhizal swards.

7.3 SYMBIOSIS AND PRIMARY SUCCESSION

'Primary succession' refers to the sequence of changes in vegetation initiated when a substratum that has never borne plants, fungi, or microorganisms is exposed. Such substrata arise, for example, on the lava flows of volcanoes, after retreat of a glacier, and on derelict land generated by mining or other industrial activity.

This section concerns the contribution of three groups of symbioses (lichens, nitrogen-fixing associations in plants, and mycorrhizas) to the colonization of barren substrata and to the subsequent sequence of vegetational change. (These sequences are known as *seres*.)

7.3.1 **Lichens**

Lichens are usually the first colonists of bare rock (unless the rock is physically unstable), and lichens and mosses commonly dominate these substrata for hundreds of years. Vascular plants usually colonize only when sufficient soil has developed, through weathering of the rock (promoted by acids released from lichens) and capture of wind-blown particles in cracks.

7.3.2 Plants with nitrogen-fixing bacteria

If the newly exposed surface has substantial amounts of fine particles (e.g. sand, silt, clay), the first colonists are not lichens but vascular plants. The substratum usually contains very little nitrogen, which is the principal plant nutrient absent from most rock minerals. Two predictions arise:

(a) Nitrogen-fixing plants are the first colonists of the barren substratum (i.e. they are pioneer species). This is because their requirements for combined nitrogen are very low.

(b) Nitrogen-fixing plants facilitate succession. The litter from nitrogen fixing plants would increase the nitrogen content of the soil, so promoting invasion by plants with higher nitrogen requirements.

There is considerable evidence that nitrogen-fixing plants can facilitate succession (prediction (b)), and this issue will be considered first. The general importance of nitrogen-fixing plants as pioneers in primary seres (prediction (a)) is less certain, as is discussed in section 7.3.4.

7.3.3 Nitrogen-fixing plants facilitate succession

Nitrogen-fixing plants in a variety of primary seres have been demonstrated to increase soil nitrogen and to promote invasion by other plants. Three examples are considered here.

(a) *Nitrogen-fixing lupins and the reclamation of derelict land*
The tree lupin *Lupinus arboreus* is widely used in programmes to reclaim derelict land. For example, it grows well on waste tips of disused china-clay works in Cornwall UK, with a substratum of sand and mica containing virtually no nitrogen. The symbiotic rhizobia in the lupins fix c. 180 kg N ha^{-1} year^{-1}, and nitrogen from the plant litter accumulates in the soil in a readily mineralizable form. Within five years of introducing lupins, grasses grow vigorously on the china-clay sites, and in ten years shrubs and trees have invaded. Within a decade, the barren substratum is transformed into a self-sustaining shrub community (Bradshaw and Chadwick 1980).

(b) *Nitrogen-fixing plants and vegetational changes at Glacier Bay, Alaska*
At Glacier Bay, Alaska, glaciers have retreated by 100 km over the last 220

years, and three distinctive plant communities can be recognized at increasing distance from the present glacial front: a pioneer community of *Salix*, *Dryas*, and *Epilobium*; a shrub community of *Alnus*, *Salix*, and *Populus*, often known as an alder thicket; and, furthest from the glacier, conifer forest, dominated by *Picea sitchensis*. Both the *Dryas* in the pioneer community and *Alnus* in the shrub community are actinorhizal (i.e. they contain nitrogen-fixing *Frankia*) (Lawrence *et al.* 1967).

The nitrogen content of the soil increases progressively with time since the substratum is colonized by plants (Fig. 7.2), and this effect can be attributed primarily to the nitrogen-fixing plants. *Alnus* may be of particular importance, as is indicated by the growth form of *Salix*. When associated with *Dryas* in the pioneer community, *Salix* is prostrate and its leaves are yellow, indicative of nitrogen limitation; but *Salix* plants associated with *Alnus* are erect and have green leaves.

(c) *Nitrogen-fixing* Myrica *and the pattern of succession after the eruption of the volcano Kilauea Iki on Hawaii*

There are no native nitrogen-fixing plants on Hawaii, but several nitrogen-fixing species have been introduced by man. These include an actinorhizal shrub, *Myrica faya*, brought to the islands by the Portuguese in the 1800's.

The volcano Kilauea Iki on Hawaii Island erupted in 1959, and some (but not all) of the affected region has subsequently been colonized by *Myrica faya*. It has therefore been possible to examine directly the effect of a nitrogen-fixing plant on the rate and pattern of succession (Vitousek and Walker 1989).

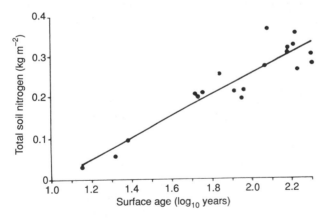

Fig. 7.2 Variation in soil nitrogen content with distance from the glacier front at Glacier Bay, Alaska. (From J. Miles (1979). *Vegetation dyanamics*. Chapman and Hall, London.)

Myrica fixes nitrogen at a rate of 10–20 kg ha^{-1} year^{-1}, and the soil around this plant is greatly enriched in nitrogen (Table 7.2(a)). In areas of the volcanic ash that are not colonized by *Myrica*, the dominant plant is the native shrub *Metrosideros polymorpha*, and the net input of nitrogen to the soil is *c*. 0.2 kg ha^{-1} year^{-1}. In glasshouse experiments, soil obtained from the environs of *Myrica* bushes promotes the growth of both *Metrosideros polymorpha* and an aggressive introduced species, *Psidium cattleianum* (the strawberry guava) (Table 7.2(b)). Vitousek and Walker (1989) suggest that nitrogen enrichment of the soil by *Myrica* may result in more rapid revegetation of the volcanic ash and colonization by a wide range of plant species. They express concern that *Myrica* may facilitate invasion by introduced species, at the expense of Hawaiian endemics.

Table 7.2 The nitrogen-fixing symbiosis in *Myrica faya* and soil fertility in Hawaii Volcanoes National Park
(a) Concentration of ammonia in soil under plants at three sites in the National Park. (For details of locations, see Vitousek and Walker (1989).) Mean values of ammonia content were significantly different ($p < 0.05$)

Plant	Ammonia content ($\mu g\, NH_3-N\, g^{-1}$ soil) (mean ± s.e.)		
	Upper Byron's Ledge	Lower Byron's Ledge	Devastation Trail
Myrica faya (N-fixer)	2.9 ± 0.8	2.4 ± 0.6	12.6 ± 4.7
Metrosideros polymoryha (non-fixer)	2.2 ± 0.8	0.6 ± 0.2	1.9 ± 0.7

(b) Growth of plants raised under glasshouse conditions, in soils collected from under *Metrosideros* and *Myrica* plants at Upper Byron's Ledge

Yields of both *M. polymorpha* and *P. cattleianum* were enhanced significantly ($p < 0.01$) by soil from environs of the N-fixer *M. faya*.

Source of soil	Total dry weight (g) per plant (mean ± s.e.)	
	Metrosideros polymorpha	*Psidium cattleianum*
Myrica faya	0.99 ± 0.15	1.12 ± 0.12
Metrosideros polymorpha	0.73 ± 0.06	0.58 ± 0.02

(Data from Vitousek and Walker (1989)).

It is of note that vegetation does develop (albeit slowly) on the volcanic ash in the absence of *Myrica*. Nitrogen-fixing plants may promote succession, but they are not required for the revegetation of barren substrata. In fact, nitrogen-fixers are absent from all successional stages in some seres, e.g. mudflows and scree from disuses slate mines (Vitousek *et al.* 1989).

7.3.4 Nitrogen-fixing plants rarely dominate the pioneer community in primary seres

Although nitrogen-fixing plants promote succession, the prediction made in section 7.3.2(a) that these plants are particularly important as first colonists of newly exposed substrata is not generally valid (Grubb 1986; Vitousek *et al.* 1989). Let us consider some examples.

The actinorhizal *Dryas* is a major member of the pioneer community at Glacier Bay (see above). *Dryas*, however, fails to colonize many sites exposed by glacial retreat, where the soil is either waterlogged or acidic. Yet the pattern and pace of subsequent vegetational changes at these sites are comparable to that at Glacier Bay; the pioneer communities lacking *Dryas* are replaced by alder thickets, and then conifer forests.

After the eruption of Mount St Helens volcano in 1980, some of the first plant colonists were the lupins *Lupinus latifolius* and *L. lepidus*. As with the lupins in the china-clay pits of Cornwall (see above), the lupins on Mount St Helens increased the soil nitrogen in their immediate environs, and promoted colonization by the other plants. The lupins on Mount St Helens did not, however, disperse over large distances, and their effect was very localized. For most of the landscape, the input from nitrogen fixation was trivial, and the principal sources of nitrogen in the early successional stages were arthropods blown on to the volcanic ash and rainfall (del Moral and Bliss 1993).

Nitrogen-fixing plants were entirely absent from the pioneer community after the eruption of Krakatau, Indonesia in 1883. Historical records indicate that plants first colonized the volcanic ash within six years, but it took a further 30 years for the coastal woodlands of actinorhizal *Casuarina* to develop (Whittaker *et al.* 1989).

Why are nitrogen-fixers usually not important in early stages of primary succession, when the soils are very low in nitrogen? For legumes, a major contributory factor may be heavy seeds, which are not transported over large distances by wind, and usually require animals for dispersal. The limited seed dispersal reduces both the chance that legumes colonize a newly exposed site and their ability to spread through a site (Grubb 1986). The lupins on Mount St Helens illustrate this point. This argument does not, however, apply to actinorhizal plants, most of which have small seeds (A.H. Fitter, pers. comm.).

A second constraint (applicable to both legumes and actinorhizal plants) may be that the nitrogen-fixing symbionts are not carried with the seed. The rhizobia or *Frankia* must be dispersed to new sites independently of the plant seeds, and must also survive in the physically harsh conditions in developing soils.

7.3.5 Mycorrhizas and plant succession

Mycorrhizal fungi are entirely absent from substrata generated by volcanic activity, glacial retreat, etc. The fungal propagules can be transported to the newly exposed sites by wind, water, or animals, either associated with plant material or as isolated fungal spores (Allen 1991; Koske and Gemma 1990).

The incidence of mycorrhizas in pioneer plant communities varies widely between sites. For example, mycorrhizas were entirely absent from the roots of the first plant colonists at Mount St Helens, and seven years later, only one species had formed mycorrhizas (Allen 1991); but at volcanic sites on Hawaii, 57 per cent of plant species were mycorrhizal within eight years of colonization (Gemma and Koske 1990). Most of the pioneers at Mount St Helens were capable of forming mycorrhizas, which suggests that the incidence of mycorrhizas at this site was limited by dispersal of the fungus.

The effect of mycorrhizal fungi on early stages in succession is also very variable. This is illustrated by two contrasting studies of Allen and Allen: the first conducted on sand dunes, where mycorrhizal fungi promoted succession, and the second on semi-arid derelict land exposed to high winds, where mycorrhizal fungi retarded colonization and succession.

An early colonist of sand dunes in north America is the annual *Salsola kali*, which is gradually displaced by perennial grasses. *Salsola* does not form mycorrhizas, and it grows very poorly if the soil contains VA-mycorrhizal fungi. The perennial grasses are mycorrhizal. Allen and Allen (1984) found that *Salsola* grew vigorously on soils lacking mycorrhizal fungi, outcompeting seedlings of the grasses *Agropyron smithii* and *Bouteloua gracilis*. The competitive advantage was reversed on soils containing mycorrhizal fungi, with the result that the grasses became established more rapidly and the succession from annual weeds to perennial grasses was promoted.

The first colonists of the exposed site studied by Allen and Allen (1988) were usually ephemeral plants. The litter from these plants acted as a wind-break, enabling other plants, especially grasses, to thrive. The ephemerals (like *Salsola* on the sand dunes) grew poorly in soils bearing mycorrhizal fungi, but other plants colonizing the mycorrhizal soil lacked protection and established slowly. In these harsh conditions, mycorrhizal fungi slowed the pace of revegetation.

Let us now consider the influence of mycorrhizal fungi on later stages in succession. Most soils with established vegetation have mycorrhizal fungi. If these fungi can form associations with relatively few plant species, then they would promote the persistence of those species, and inhibit invasion by other plants that have different mycorrhizal fungi. For example, the ericoid mycorrhizal fungi associated with *Calluna* (heather) is believed to inhibit the growth of ectomycorrhizal fungi, and so retard invasion of heathland by ectomycorrhizal conifers (Robinson 1972). However, most mycorrhizal fungi have a very broad plant range (see Chapter 5, section 3.2), and they would not be expected to retard succession in this way.

7.4 THE ECOLOGICAL IMPACT OF PLANT SYMBIOSES ON VEGETATIONAL PROCESSES: A SUMMARY

Plants with nitrogen-fixing symbioses tend to increase the nitrogen content of soils. This property is used both in land reclamation and as a nitrogen fertilizer in many agricultural practices (e.g. legumes in crop rotation in northern Europe, the *Azolla*-cyanobacterial symbiosis in rice production in south-east Asia). In natural ecosystems, the input of nitrogen to soils from nitrogen-fixing plants may both accelerate succession and change the pattern of succession. However, nitrogen-fixers are rarely pioneers in natural systems (probably because of the limited dispersal of their seeds), and some primary seres lack any successional stage dominated by nitrogen-fixing plants.

Mycorrhizal fungi promote the uptake by plants of phosphate and ammonium ions from soils, and this can contribute to the tight cycling of N and P in terrestrial ecosystems. Mycorrhizal fungi can also influence the competitive interactions between plants, either (a) by altering the competitive balance between plant species, which all form associations with the mycorrhizal fungus; or (b) by promoting the vigour of plants with which they associate, while being detrimental to other plants.

As a consequence, the effect of mycorrhizas on plant communities is variable. They may promote plant species diversity (by reducing the competitive superiority of the dominant species); they may facilitate or retard early stages in succession; and they may influence the pace of later successional stages.

8
Symbiosis and the eukaryotic cell

Eukaryotes have acquired the capacity to respire aerobically and to photosynthesize, exclusively through symbiosis; and all symbionts with aerobic respiration and many photosynthetic symbionts have evolved into organelles, the mitochondria and plastids respectively. This chapter concerns two issues: the processes underlying the transformation of intracellular symbionts into organelles, and the identity of the microorganisms from which mitochondria and plastids evolved.

The possibility that mitochondria and plastids are derived from symbiotic microorganisms has been raised repeatedly since the organelles were first recognized in the nineteenth century. Proponents of this view were ignored or denigrated by most contemporary biologists. Current acceptance of the symbiotic origins of these organelles is a direct result of, first, the persuasive advocacy of Lynn Margulis (1970, 1993) and, second, the advent of molecular techniques, which provided the means of testing the hypothesis.

The evidence that mitochondria and plastids have evolved from intracellular symbionts is now overwhelming. These organelles have translation machinery (i.e. ribosomes, tRNAs, etc. for protein synthesis) distinct from the nucleocytoplasm, and the sequences of organellar genes, such as rRNA genes, are more closely allied with eubacteria than eukaryotes.

Margulis has argued consistently that eukaryotes have acquired three organelles through symbiosis: mitochondria, plastids, and flagella (Fig. 8.1). She considers spirochaetes to have given rise to eukaryotic flagella, from which all other microtubule systems have evolved. In other words, eukaryotes are proposed to have acquired *motility* and a *novel gene* (coding for tubulin, the constituent protein of microtubules) from symbiotic spirochaetes, but novel metabolic capabilities from mitochondria and plastids. The case for a symbiotic origin of microtubular systems is weak. In particular, there is no persuasive evidence that any spirochaete has tubulin. However, there has been a resurgence of interest in this hypothesis with the demonstration that the basal bodies of flagella in *Chlamydomonas* have DNA coding from flagellar proteins (Hall *et al.* 1989). Because a novel metabolic capability is not implicated, the evolutionary origins of microtubular systems will not be considered further in this book.

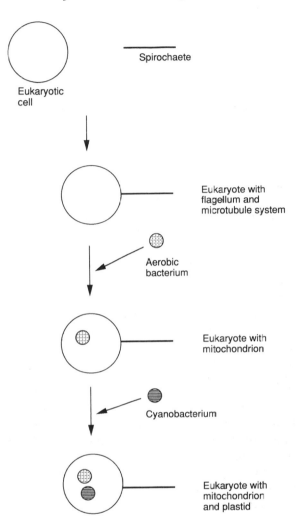

Fig. 8.1 The serial endosymbiosis theory of the origin of eukaryotic organelles. Margulis has proposed three sequential symbiotic events by which three groups of eukaryotic organelles were acquired: spirochaetes gave rise to eukaryotic flagella and microtubular systems, and Proteobacteria and cyanobacteria evolved into mitochondria and plastids respectively.

8.1 THE CHARACTERISTICS OF SYMBIONT-DERIVED ORGANELLES

8.1.1 The location of organelle-specific genes

Mitochondria and plastids differ from intracellular symbionts in a number of ways, but the crucial difference is that genes essential to the function of organelles (but not symbionts) are located in the nucleus of the host cell. Gray (1989) has estimated that between 80 and 90 per cent of proteins specific to mitochondria and plastids are coded in the nucleus. Even the genes for subunits of a single protein may have different subcellular locations. For example, in plants the small subunit of ribulose bisphosphate carboxylase (RuBisCo, the enzyme that fixes carbon dioxide in plastids) is coded in the nucleus, and the large subunit is coded in the plastid. In mammals, three of the eight subunits of cytochrome oxidase (a component of the respiratory electron transport chain in mitochondria) are coded in the mitochondria and the remainder in the nucleus.

It is agreed that the various organelle-specific genes in the nucleus have been transferred from the organelle (Fig. 8.2). This characteristic can be used to distinguish between intracellular symbionts and symbiont-derived organelles (Douglas 1992).

8.1.2 The DNA content of organelles

In addition to having transferred genes to the host nucleus, organelles have lost substantial amounts of DNA coding for functions duplicated by the host nucleus. The genome size of both mitochondria (14–2400 kbp) and plastids (120–200 kbp) is substantially smaller than that of bacteria (for example, the DNA content of *Escherichia coli* is 4400 kbp).

The only microbial symbionts known to have a very low DNA content are the intracellular cyanobacteria in *Cyanophora paradoxa*, with a genome size of 127 kbp (Wassman *et al.* 1987). These cyanobacteria cannot be considered as organelles because gene transfer to the host has not been demonstrated; in particular, both subunits of RuBisCo are coded by the cyanobacterium (Starnes *et al.* 1985).

8.2 THE HOSTS IN WHICH MITOCHONDRIA AND PLASTIDS EVOLVED

There is an erroneous belief, frequently expressed in the literature, that the first hosts of mitochondria were bacteria (either eubacteria or

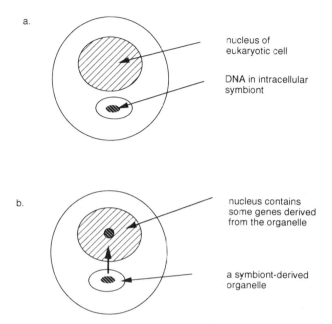

a.

nucleus of
eukaryotic cell

DNA in intracellular
symbiont

b.

nucleus contains
some genes derived
from the organelle

a symbiont-derived
organelle

Fig. 8.2 The distinction between intracellular symbionts and symbiont-derived organelles. An intracellular symbiont can be considered to have evolved into an organelle if genes essential to its function have been transferred to the eukaryotic cell nucleus. (a) An intracellular symbiont with a genome distinct from that of the host cell nucleus. (b) A symbiont-derived organelle: the nucleus of the eukaryotic cell contains some genes derived from the organelle.

archaebacteria), and that acquisition of mitochondria marked the origin of eukaryotes. Arising from this fallacy is the claim that eukaryotes evolved when the Earth's atmosphere became oxic, about 2×10^9 years ago. In fact, several groups of protists (unicellular eukaryotes), including metamonads, microsporidians, and diplomonads, diverged before mitochondria were acquired (Sleigh 1989), and today *c.* 1000 species of protists which primitively lack mitochondria are known (Cavalier-Smith 1987*a*). Eukaryotes evolved under anoxic conditions, and later came to exploit oxygen in aerobic respiration, through symbiosis. The first putative eukaryotic hosts of mitochondria are, however, obscure.

Plastids have evolved in several groups of protists, including dinoflagellates, euglenoids, and heterokonts (but the immediate ancestor of the chlorophytes, from which the terrestrial plants evolved, is uncertain) (Sleigh 1989). From phylogenetic analyses of protists, Perasso *et al.* (1989) have estimated that most plastids were acquired in the late Precambrian era, probably between 0.8 and 0.6×10^9 years ago.

Let us return to the fundamental characteristic of symbiont-derived organelles, that genes essential to their function are located in the host nucleus. The nucleus most likely to acquire symbiont DNA is that of the cell housing the symbionts. Organelles can therefore be expected to evolve in hosts in which the cells containing the symbionts also give rise to host progeny. This is the norm among the unicellular protists. In most multicellular eukaryotes, the cells containing symbionts are distinct from the host's reproductive propagules, and therefore the symbionts of multicellular hosts are very unlikely to evolve into organelles.

Consistent with this line of reasoning, all known symbiont-derived organelles have arisen in protist hosts. This may be a major constraint on the variety (in metabolic terms) of organelles. In particular, the usual hosts of chemosynthetic and nitrogen-fixing symbionts are animals and plants respectively (see Tables 2.4 and 2.5). Nitrogen-fixing and chemosynthetic organelles are, consequently, very unlikely to evolve.

8.3 HOW SYMBIONTS EVOLVE INTO ORGANELLES

The transfer of DNA from intracellular symbionts/organelles to the nucleus is not an improbable event. DNA in eukaryotic cells is relatively mobile, and there are a number of reports of DNA sequences that have been transferred between the nucleus, mitochondria and plastids (Gray 1989).

In the evolution of organelles, the net movement of genes is asymmetric, with the nucleus as the usual recipient and symbionts/organelles as donors. Three features of nuclei may contribute to this asymmetry.

(1) The nucleus can 'scavenge' free cytoplasmic DNA which may result from the lysis of an intracellular symbiont or organelle (Thorsness and Fox 1990).

(2) The nuclear genome is much less compact than bacterial and organellar genomes, and therefore incoming DNA is less likely to cause disruption by integrating into a region coding for essential functions in the nucleus than organelles.

(3) Most eukaryotic cells have one nucleus but many organelles, and therefore a gene acquired by the nucleus is far more likely to be transmitted to daughter cells than one acquired by a single organelle.

Once an organelle-specific gene is incorporated into the nuclear genome, it is translated in the cytoplasm (and not its original location within the symbiont/organelle). Consequently, for the nuclear-encoded gene to be functional, the protein must be targeted back to the appropriate organelle. Cavalier-Smith (1987*b*) has argued that the evolution of faithful protein-targeting may be a far greater barrier to the evolution of organelles than the transfer of genes to the host nucleus.

8.4 PLASTID ORIGINS

The traditional view is that plastids have evolved three times, and the three groups can be distinguished by their chlorophyll pigments:

(1) Plastids with chlorophyll *a* and *b* in the Chlorophyta (green algae), Euglenophyta, and the land plants. These were believed to have evolved from prochlorons (photosynthetic bacteria with both chlorophylls *a* and *b*).

(2) Plastids with chlorophyll *a* and phycobilisomes in the Rhodophyta (the red algae), derived from cyanobacteria.

(3) Plastids with chlorophyll *a* and *c* in Chromophyta (including the brown seaweeds and diatoms). The putative bacterial ancestor of these plastids is unknown.

Recent molecular analyses do not support this view. It is becoming increasingly clear that all plastids arose from a very restricted group of cyanobacteria, and none are closely related to prochlorons (Howe *et al.* 1992). Plastids may be monophyletic (S.E. Douglas and Turner 1991) or they may have evolved twice, one lineage giving rise to the plastids in the Chlorophyta and land plants and the other lineage to all other plastids (Howe *et al.* 1992).

The evolutionary origins of plastids is complicated by the fact that many were acquired not as cyanobacteria but as symbionts/organelles already established within protist hosts. The principal evidence that some plastids are products of two successive symbioses is structural; certain plastids have more than two bounding membranes (Whatley and Whatley 1983). As illustrated in Fig. 8.3, a plastid with four bounding membranes could arise if a eukaryote (host-I in the figure) with a cyanobacterial symbiont were acquired as an intracellular symbiont by a second eukaryote (host-II). In this condition, genes from the photosynthetic symbiont would be transferred to the nucleus of host-II, and the nucleocytoplasm of host-I, being redundant, would be reduced progressively to an 'empty' space between membranes-2 and membrane-3 of Fig. 8.3(b). This is the condition of plastids in the Chromophyta. The plastids in dinoflagellates have just three bounding membranes, which probably arose by loss of the outermost membrane (the symbiosome membrane of host-II).

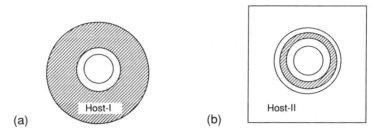

(a)   (b)

Fig. 8.3 The origin of complex plastids. (a) A cyanobacterium is acquired as an intracellular symbiont by a eukaryote (host-I). The nucleocytoplasm of host-I is shaded. 'Simple' plastids are derived from this single symbiosis. (b) Host-I and the enclosed cyanobacterium are acquired by a second eukaryote (host-II). In the complex four-membrane bound plastid derived from this association, the nucleo-cytoplasm of host-I is indicated by the shaded area and the outermost of the four membranes is the symbiosome membrane of host-II.

Let us now consider complex plastids arising when host-II acquires a eukaryote with a plastid, so that genes essential to plastid function are located in the nucleus of host-I. In this condition, a functional nucleo-cytoplasm of host-I is retained. This has occurred in at least one protist group, the Cryptophyta, the ancestor of which acquired an unicellular red alga (McFadden 1992). Within the cryptophyte plastid, the nucleocytoplasm of the red alga is retained in a much reduced form. The red alga's nucleus, known as the nucleomorph, contains *c.* 660 kbp DNA distributed between three linear chromosomes (Fig. 8.4).

8.5 THE EVOLUTIONARY ORIGINS OF MITOCHONDRIA

The DNA sequence of ribosomal RNA genes is very useful in assessing the evolutionary origins of mitochondria because mitochondria, the nucleo-cytoplasm of eukaryotes, and all bacteria have these genes. Most studies of rDNA sequences assign mitochondria to the alpha-Proteobacteria (e.g. Cedergren *et al.* 1988). This is consistent with biochemical data; for example, John and Whatley (1975) have reported the close similarity between the electron transport chains of mitochondria and the alpha-Proteobacterium *Paracoccus denitrificans*. It is uncertain whether the mitochondria have one or multiple origins within the alpha-Proteobacteria (Gray *et al.* 1989).

A singular feature of mitochondria is the apparent absence of analogous symbionts. There are no known symbioses between protists lacking mitochondria and aerobically respiring bacteria. (An amoeba, *Pelomyxa*

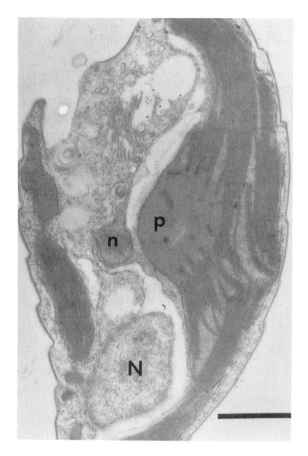

Fig. 8.4 The nucleomorph in the complex plastids of cryptophytes. A longitudinal section through the cryptophyte *Komma caudata* showing the main nucleus (N) and the nucleomorph (n) associated with the pyrenoid (p) and plastid. Scale: 1 μm. (Reproduced from McFadden (1992))

palustris, was once thought to contain this type of symbiosis, but the symbiotic bacteria are now known to be methanogens (van Bruggen *et al*. 1988). It is unclear why, of all metabolic capabilities, aerobic respiration is not used through symbiosis today. Perhaps such associations are inherently improbable (but why?). Alternatively, any association may be out-competed by eukaryotes with mitochondria.

8.6 CONCLUSIONS

The foundation of this text is that symbiosis has expanded the metabolic repertoire of eukaryotes. The greater part of the book has concerned associations that evolved in the Phanerozoic; the past 600 million years and the age of the multicellular eukaryotes—the animals, plants, and fungi. In an evolutionary sense, however, the most significant symbiosis in eukaryotes is Precambrian in origin: the acquisition of aerobic respiration through mitochondria. It is not entirely fanciful to suggest that, without this symbiosis, the eukaryotes would today be relegated to a few anaerobic environments, and the world would have been dominated by the eubacteria.

References

Albert, M. J., Malthan, V. I., and Baker, S. J. (1980). Vitamin B_{12} synthesis by human small intestine bacteria. *Nature*, **283**, 781–2.

Alexander, M. (1985). Ecological constraints on nitrogen fixation in agricultural ecosystems. *Advances in Microbial Ecology*, **8**, 164–183.

Allen, E. B., and Allen, M. F. (1984). Competition between plants of different successional stages: mycorrhizae as regulators. *Canadian Journal of Botany*, **62**, 2625–9.

Allen, E. B., and Allen, M. F. (1988). Facilitation of succession by the nonmycotrophic coloniser *Salsola kali* (Chemopodiaceae) on a harsh site: effects on mycorrhizal fungi. *American Journal of Botany*, **75**, 257–66.

Allen, M. F. (1991). *The ecology of mycorrhizae*. Cambridge University Press.

Appleby, C. A. (1984). Leghaemoglobin in legumes. *Annual Review of Plant Physiology*, **35**, 443–68.

Appleby, C. A., Bogusz, D., Dennis, E. S., and Peacock, W. J. (1988). A role for haemoglobin in all plant roots? *Plant Cell and Environment*, **11**, 359–67.

Bauchop, T. (1977). Foregut fermentation. In *Microbial ecology of the gut* (ed. R. T. J. Clarke and T. Bauchop), pp. 223–51. Academic Press, London.

Bauchop, T. (1989). Biology of gut anaerobic fungi. *BioSystems*, **23**, 53–64.

Becard, G., and Piche, Y. (1989). Fungal growth stimulation by carbon dioxide and root exudates in vesicular-arbuscular mycorrhizal symbiosis. *Applied and environmental Microbiology*, **55**, 2320–25.

Bird, S. H., and Lang, R. A. (1978). The effects of defaunation of the rumen on the growth of cattle on low protein–high energy diets. *British Journal of Nutrition*, **40**, 163–7.

Blumwald, E., Fortin, M. G., Rea, P. A., Verma, D. P. S., and Poole, R. J. (1985). Presence of host-plasma membrane type H^+-ATPase in the membrane envelope enclosing the bacteroids in soybean root nodules. *Plant Physiology*, **78**, 665–72.

Bohatier, J., Senaud, J. and Benyahya, M. (1990) In situ degradation of cellulose fibres by the entodinionorph rumen ciliate, *Polyplastron multivesiculatum*. *Protoplasma* **154**, 122–31.

Boucher, D. H. (1985). The idea of mutualism, past and present. In *The biology of mutualism* (ed. D. H. Boucher), pp. 1–28. Croom Helm, London.

Bradley, D. J., Butcher, G. W., Galfre, G., Wood, E. A., and Brewin, N. J. (1985). Physical association between the peribacteroid membrane and lipopolysaccharide from the bacteroid outer membrane in *Rhizobium*-infected pea root nodule cells. *Journal of Cell Science*, **85**, 47–61.

Bradshaw, A. D., and Chadwick, M. J. (1980). *The restoration of land: the ecology and reclamation of derelict and degraded land*. Blackwell Scientific Publications, Oxford.

Brewin, N. J. (1992). Development of the legume root nodule. *Annual Reviews of Cell Biology*, **7**, 191–226.

Breznak, J. A. (1982). Intestinal microbiota of termites and other xylophagous insects. *Annual Review of Microbiology*, **36**, 323–43.

Brodo, I. M., and Richardson, D. H. S. (1978). Chimeroid associations in the genus *Peltigera*. *Lichenologist*, **10**, 157–70.

Brooks, M. A., and Richards, A. G. (1955). Intracellular symbiosis in cockroaches. II. Mitotic division of mycetocytes. *Science*, **122**, 242.

Brundrett, M. (1991). Mycorrhizas in natural ecosystems. *Advances in Ecological Research*, **21**, 171–315.

Buchner, P. (1965). *Endosymbioses of animals with plant microorganisms*. Wiley, London.

Cavalier-Smith, T. (1987*a*). Eukaryotes with no mitochondria. *Nature*, **326**, 332–3.

Cavalier-Smith, T. (1987*b*). The origin of eukaryote and archaeobacterial cells. *Annals of the New York Academy of Sciences*, **503**, 17–54.

Cavanaugh, C. M. (1992). Methanotroph-invertebrate symbioses in the marine environment: ultrastructural, biochemical, and molecular studies. In *Microbial growth on C_1 compounds* (ed J. C. Murrell and D. P. Kelly), pp. 315–28. Intercept Ltd., Andover, UK.

Cavanaugh, C. M., Gardiner, S. L., Jones, M. L., Jannasch, H. W., and Waterbury, J. B. (1981). Prokaryotic cells in the hydrothermal vent tube worm *Riftia pachyptila* Jones: possible chemoautotrophic symbionts. *Science*, **213**, 340–2.

Cedergren, R, Gray, M. W., Abel, Y., and Sankoff, D. (1988). The evolutionary relationships among known life forms. *Journal of Molecular Evolution*, **28**, 98–112.

Cernichiari, E., Muscatine, L., and Smith, D. C. (1969). Maltose excretion by the symbiotic algae of *Hydra viridis*. *Proceedings of the Royal Society of London B*, **173**, 557–76.

Charnley, A. K., Hunt, J., and Dillon, R. J. (1985). The germ-free culture of desert locusts, *Schistocerca gregaria*. *Journal of Insect Physiology*, **31**, 477–85.

Chesnik, J. M., and Cox, E. R. (1987). Synchronised sexuality of an algal symbiont and its dinoflagellate host, *Peridinium balticum* (Levander) Lemmermann. *BioSystems*, **21**, 69–78.

Clark, M. A., Baumann, L., Munson, M. A., Baumann, P., Campbell, B. C., Duffus, J. E. Osborne, L. S., and Moran, N. A. (1922). The eubacterial endosymbionts of whiteflies (Homoptera: Aleyrodoidea) constitute a lineage distinct from the endosymbionts of aphids and mealybugs. *Current Microbiology*, **25**, 119–23.

Day, D. A., and Copeland, L. (1991). Carbon metabolism and compartmentation in nitrogen-fixing legume nodules. *Plant Physiology and Biochemistry*, **29**, 185–201.

Day, D. A., Yang, L. J. O., and Udvardi, M. K. (1990). Nutrient exchange across the peribacteroid membrane of isolated symbiosomes. In *Nitrogen fixation: achievements and objectives* (ed. P. M. Gresshoff, G. Stacey, and W. E. Newton), pp. 219–26. Chapman and Hall, New York.

de Philip, P., Batut, J., and Bristard, P. (1990). *Rhizobium meliloti* FixL is an oxygen sensor and regulates *R. meliloti nifA* and *fixK* genes differently in *Escherichia coli. Journal of Bacteriology*, **172**, 4255–62.

del Morel, R., and Bliss, L. C. (1993). Mechanisms of primary succession: insights resulting from the eruption of Mount St Helens. *Advances in Ecological Research*, **24**, 1–66.

Distel, D. L. (1990). Detection, identification and phylogenetic analysis of endosymbiotic bacteria using ribosomal RNA sequences. *Endocytobiology*, **4**, 339–42.

Distel, D. L., Lane, D. J., Olsen, G. J., Giovannoni, S. J., Pace, B., Pace, N. R., Stahl, D. A., and Felbeck, H. (1988). Sulfur-oxidizing bacterial endosymbionts: analysis of phylogeny and specificity by 16S rRNA sequences. *Journal of Bacteriology* **170**, 2506–10.

Douglas, A. E., (1983). Establishment of the symbiosis in *Convoluta roscoffensis*. *Journal of the marine biological Association UK*, **63**, 419–34.

Douglas, A. E. (1989). Mycetocyte symbiosis in insects. *Biological Reviews*, **64**, 409–34.

Douglas, A. E. (1992), Symbiosis in evolution. *Oxford Surveys in Evolutionary Biology*, **9**, 349–82. Oxford University Press.

Douglas, A. E., and Dixon, A. F. G. (1987). The mycetocyte symbiosis of aphids: variation with age and morph in virginoparae of *Megoura viciae* and *Acyrthosiphon pisum*. *Journal of Insect Physiology*, **33**, 109–13.

Douglas, A. E., and Smith, D. C. (1983). The costs of symbionts to the host in green hydra. *Endocytobiology*, **2**, 631–47.

Douglas, A. E., and Smith D. C. (1984). The green hydra symbiosis. VIII. Mechanisms in symbiont regulation. *Proceedings of the Royal Society of London B*, **221**, 291–319.

Douglas, S. E., and Turner, S. (1991). Molecular evidence for the origin of plastids from a cyanobacterium-like ancestor. *Journal of Molecular Evolution*, **33**, 267–73.

Dumment, M. W., and van Soest, P. J. (1985). A nutritional explanation for body size patterns of ruminant and nonruminant herbivores. *American Naturalist*, **125**, 641–72.

Dunlap, P. V. (1984). Physiological and morphological state of the symbiotic bacteria from the light organs of ponyfish. *Biological Bulletin*, **167**, 410–25.

Eardly, B. D., Young J. P. W., and Selander, R. K. (1992). Phylogenetic position of *Rhizobium sp.* strain Or191, a symbiont of both *Medicago sativa* and *Phaseolus vulgaris*, based on partial sequences of the 16S rRNA and *nifH* genes. *Applied and Environmental Microbiology*, **58**, 1809–15.

Eberle, M. W., and McLean, D. L. (1983). Observation of symbiote migration in human body lice with scanning and transmission electron microscopy. *Canadian Journal of Microbiology*, **29**, 755–62.

Eisen, J. A., Smith, S. W., and Cavanaugh, C. M. (1992). Phylogenetic relationships of chemoautotrophic bacterial symbionts of *Solemya velum* Say (Mollusca: Bivalvia) determined by 16S rRNA gene sequence analysis. *Journal of Bacteriology*, **174**, 3416–21.

Farrar, J. F. (1976). Ecological physiology of the lichen *Hypogymnia physodes*. II. Effects of wetting and drying cycles and the concept of physiological buffering. *New Phytologist*, **77**, 105–13.

Fenchel, T., and Finlay, B. J. (1991a). The biology of free-living anaerobic ciliates. *European Journal of Protistology*, **26**, 201–15.

Fenchel, T., and Finlay, B. J. (1991b). Synchronous division of an endosymbiotic methanogenic bacterium in the anaerobic ciliate *Plagiopyla frontata* Kahl. *Journal of Protozoology*, **38**, 22–8.

Finan, T. M., Wood, J. M., and Jordan, D. C. (1983). Symbiotic properties of C_4

dicarboxylic acid transport mutants of *Rhizobium leguminosarum*. *Journal of Bacteriology*, **154**, 1403–13.

Fisher, C. R. (1990). Chemoautotrophic and methanotrophic symbioses in marine invertebrates. *Reviews in Aquatic Sciences*, **2**, 399–436. CRC Press, Florida, USA.

Fisher, R. F., and Long, S. R. (1992). *Rhizobium*-legume signal exchange. *Nature*, **357**, 655–60.

Fitt, W. K., and Cook, C. B. (1990). Some effects of host feeding on growth of zooxanthellae in the marine hydroid *Myrionema ambionense* in the laboratory and in nature. *Endocytobiology*, **4**, 281–4.

Fitter, A. H. (1977). Influence of mycorrhizal infection on competition for phosphorus and potassium by two grasses. *New Phytologist*, **79**, 119–25.

Frost, T. M., and Williamson, C. E. (1980). In situ determination of the effect of symbiotic algae on the growth of the freshwater sponges *Spongilla lacustris*. *Ecology*, **6**, 1361–70.

Futuyma, D. J., and Moreno, G. (1988). The evolution of ecological specialisation. *Annual Review of Ecology and Systematics*, **19**, 207–33.

Gadgil, R. L. and Gadgil, P. D. (1971). Mycorrhiza and litter decomposition. *Nature*, **233**, 133.

Gemma, J. N., and Koske, R. E. (1990). Mycorrhizae in recent volcanic substrates in Hawaii. *American Journal of Botany*, **77**, 1193–200.

Giovannoni, S. J., Britschgi, T., Moyer, C. J., and Field, K. G. (1990). Genetic diversity in Sargasso Sea bacterioplankton. *Nature*, **345**, 60–3.

Grajal, A., Stahl, S. D., Parra, R., Dominguez, M. G., and Neher, A. (1989). Foregut fermentation in the hoatzin, a neotrophical leaf-eating bird. *Science*, **245**, 1236–8.

Gray, M. W. (1989). Origin and evolution of mitochondrial DNA. *Annual Review of Cell Biology*, **5**, 25–50.

Gray, M. W., Cedergren, R., Abel, Y., and Sankoff, D. 1989. On the evolutionary origin of the plant mitochondrion and its genome. *Proceedings of the National Academy of Sciences of USA*, **86**, 2267–71.

Greenhalgh, G. N. and Anglesea, D. (1979). The distribution of algal cells in lichen thalli. *Lichenologist*, **11**, 283–92.

Grime, J. P., Mackay, J. M. L., Hillier, S. H., and Read, D. J., (1987). Floristic diversity in a model stystem using experimental microcosms. *Nature*, **328**, 420–2.

Grubb, P. J. (1986). The ecology of establishment. In *Ecology and the Design of Landscape* (ed A. J. Bradshaw, D. A. Goode, and E. Thorpe), pp. 83–97. Blackwell, Oxford.

Hall, J. L., Ramanis, Z., and Luck, D. J. L. (1989). Basal body/centriolar DNA: molecular genetic studies in *Chlamydomonas*. *Cell*, **59**, 121–32.

Harley, J. E., and Smith, S. E. (1983). *Mycorrhizal Symbiosis*. Academic Press, London.

Harmer, R., and Alexander, I. J. (1985). Effects of root exclusion on nitrogen transformations and decomposition processes in spruce humus. In *Ecological interactions in soil* (ed. A. H. Fitter, D. J. Read, and M. B. Usher), pp. 267–77. Blackwell, Oxford.

Hastings, J. W. (1983). Biological diversity, chemical mechanisms and the evolutionary origins of bioluminescent systems. *Journal of molecular Evolution*, **19**, 309–21.

Hatcher, B. G. (1990). Coral reef primary productivity: a hierarchy of pattern and process. *Trends in Ecology and Evolution*, **5**, 149–55.

Hawksworth, D. L. (1988). Coevolution of fungi with algae and cyanobacteria in lichen symbiosis. In *Coevolution of fungi with plants and animals* (ed. K. A. Pirozynski and D. L. Hawksworth), pp. 125–48. Academic Press, London.

Hawksworth, D. L., and Hill, D. J. (1984). *The lichen-forming fungi*. Blackie, Glasgow.

Haygood, M. G. (1990). Relationship between the luminous bacterial symbiont of the Caribbean flashlight fish, *Kryptophanaron alfredi* (family Anomalopodiae) to other luminous bacteria based on bacterial luciferase (*luxA*) genes. *Archives of Microbiology*, **154**, 496–503.

Haygood, M. G., Distel, D. L., and Herring, P. J. (1992). Polymerase chain reaction and 16S rRNA gene sequences from the luminous bacterial symbionts of two deep-sea anglerfishes. *Journal of marine biological Association UK*, **72**, 149–59.

Herring, P. J. (1988). Luminescent organs. In *The Mollusca*, Vol. 11 (ed. E. R. Trueman and M. R. Clarke), pp. 27–49 Academic Press, London.

Hoegh-Guldberg, O., and Smith, G. J. (1987). Influence of the population density of zooxanthellae and supply of ammonium on the biomass and metabolic characteristics of the reef corals, *Seriatopora hystrix* and *Stylophora pistillata*. *Marine Ecology Progress Series*, **57**, 173–86.

Hoegh-Guldberg, O., McCloskey, L. R., and Muscatine, L. (1987). Expulsion of zooxanthellae by symbiotic cnidarians from the Red Sea. *Coral Reefs*, **5**, 201–4.

Hofmann, R. R. (1989). Evolutionary steps of ecophysiological adaptation and diversification of ruminants: a comparative view of their digestive system. *Oecologia*, **78**, 443–57.

Hogan, M. E., Veivers, P. C., Slaytor, M., and Czolij, R. T. (1988*a*). The site of cellulose breakdown in higher termites (*Nasutitermes walkeri* and *Nasutitermes exitosus*). *Journal of Insect Physiology*, **34**, 891–9.

Hogan, M. E., Schulz, M. W., Slaytor, M., Czolij, R. T., and O'Brien, R. W. (1988*b*). Components of the termite and protozoal cellulases of the lower termite *Coptotermes lacteus* Froggatt. *Insect Biochemistry*, **18**, 45–51.

Hohman, T. C., McNeill, P. L., and Muscatine, L. (1982). Phagosome–lysosome fusion inhibited by algal symbionts of *Hydra viridis*. *Journal of Cell Biology*, **94**, 56–63.

Honegger, R. (1984). Cytological aspects of the mycobiont–phycobiont relationship in lichens. *Lichenologist*, **16**, 111–27.

Honegger, R. (1992). Lichens: mycobiont–photobiont relationships. In *Algae and symbioses* (ed. W Reisser), pp. 255–76. Biopress Ltd, Bristol, UK.

Howe, C. J., Beanland, T. J., Larkum, A. W. D., and Lockhart, P. J. (1992). Plastid origins. *Trends in Ecology and Evolution* **7**, 378–83.

James, P. W., and Henssen, A. (1976). The morphology and taxonomic significance of cephalodia. In *Lichenology: progress and problems* (ed. D. H. Brown, D. L. Hawksworth, and R. H. Bailey), pp. 27–77. Systematics Association Special Volume 8. Academic Press, London.

John, P., and Whatley, F. R. (1975). *Paracoccus denitrificans* and the evolutionary origin of the mitochondrion. *Nature* **254**, 495–8.

Kappen, L. (1988). Ecophysiological relationships in different climatic regions. In

Handbook of lichenology, Volume II, pp. 37–100. CRC Press, Boca Raton, Florida.

Koske, R. E. and Gemma, J. N. (1990). VA-mycorrhizae in strand vegetation of Hawaii: evidence for long-distance codispersal of plants and fungi. *American Journal of Botany*, **77**, 466–74.

Koucheki, H. K., and Read, D. J. (1976). Vesicular-arbuscular mycorrhizas in natural vegetation systems. III. The relationship between infection and growth in *Festuca ovina* L. *New Phytologist* **77**, 655–66.

Langer, P. (1984). Comparative anatomy of the stomach in mammalian herbivores. *Quarterly Journal of Experimental Physiology*, **69**, 615–25.

Lawrence, D. B., Schoenike, R. E., Quispel, A., and Bond, G. (1967). The role of *Dryas drumondii* in vegetation development following ice regression at Glacier Bay, Alaska, with special reference to its nitrogen fixation by root nodules. *Journal of Ecology*, **55**, 793–813.

Leronge, P., Roche, R., Faucher, C., Maillet, F., Truchet, G., Prome, J. C., and Denarie, J. (1990). Symbiotic host-specificity of *Rhizobium meliloti* is determined by a sulphated and acylated glucosamine oligosaccharide signal. *Nature*, **344**, 781–84.

Lumpkin, T. A., and Plucknett, D. L. (1980). *Azolla*: botany, physiology, and use as a green manure. *Economic Botany*, **34**, 111–53.

McAuley, P. J. (1987). Nitrogen-limitation and amino acid metabolism of *Chlorella* symbiotic with green hydra. *Planta*, **171**, 532–8.

McAuley, P. J., (1990). Partitioning of symbiotic *Chlorella* at host cell telophase in the green hydra. *Proceedings of the Royal Society of London B*, **329**, 47–53.

McDermott, T. R., Griffiths, S. M., Vance, C. R., and Graham, P. H. (1989). Carbon metabolism in *Bradyrhizobium japonicum* bacteroids. *FEMS Microbiology Reviews*, **63**, 327–40.

McFadden, G. I. (1992). Second-hand chloroplasts: evolution of cryptomonad algae. *Advances in Botanical Research*, **19**.

McFall-Ngai, M. J. (1983). The gas bladder as a central component of the leiognathid bacterial light organ symbiosis. *American Zoologist*, **23**, 907–21.

McFall-Ngai, M. J. (1990). Luminous bacterial symbiosis in fish evolution: adaptive radiation among the leiognathid fishes. In *Symbiosis as a source of evolutionary innovation* (ed. L. Margulis and R. Fester), pp. 380–409. MIT Press, Cambridge, USA.

McFall-Ngai, M. J.. and Montgomery, M. K. (1990). The anatomy and morphology of the adult bacterial light organ of *Euprymna scolopes* Berry (Cephalopoda: Sepiolidae). *Biological Bulletin*,**179**, 332–9.

McFall-Ngai, M. J., and Ruby, E. G. (1991). Symbiont recognition and subsequent morphogenesis as early events in an animal–bacterial mutualism. *Science*, **254**, 1491–4.

McNeil, N. I. (1984). The contribution of the large intestine to the energy supplies in man. *American Journal of Clinical Nutrition*, **39**, 338–42.

Margulis, L. (1970). *Origin of eukaryotic cells*. Yale University Press, New Haven, USA.

Margulis, L. (1993). *Symbiosis in cell evolution* (2nd edition). Freeman, San Francisco, USA.

Marks, G. C., and Foster, R. C. (1973). Structure, morphogenesis and ultrastructure of ectomycorrhizae. In *Ectomycorrhizae* (ed. G.C. Marks and T. T. Kozlowski), pp. 1–41. Academic Press, New York.

Martin, M. M. (1991). The evolution of cellulose digestion in insects. *Philosophical Transactions of the Royal Society of London B*, **333**, 281–8.

Molina, R., Trappe, J. M., and Stricker, G. S. (1978). Mycorrhizal fungi associated with *Festuca* in the western United States. *Canadian Journal of Botany*, **56**, 1691–5.

Munson, M. A., Baumann, P., Clark, M. A., Baumann, L., Moran, N. A., Voegtlin, D. J., and Campbell, B. C. (1991). Evidence for the establishment of aphid–eubacterium endosymbiosis in an ancestor of four aphid families. *Journal of Bacteriology*, **173**, 6321–4.

Muscatine, L. (1990). The role of symbiotic algae in carbon and energy flux in reef corals. In *Coral Reefs* (ed. Z. Dubinsky), pp. 75–87. Elsevier, Amsterdam.

Muscatine, L., Falkowski, P., Porter, J., and Dubinsky, Z. (1984). Fate of photosynthetically-fixed carbon in light and shade-adapted colonies of the symbiotic coral *Stylophora pistillata*. *Proceedings of the Royal Society of London, B* **222**, 181–202.

Nap, J.-P., and Bisseling, T. (1990). The roots of nodulins. *Physiologia Plantarum*, **79**, 407–14.

Nierzwicki-Bauer, S. A., and Haselkorn, R. (1986). Differences in mRNA levels in *Anabaena* living freely or in symbiotic association with *Azolla*. *EMBO Journal*, **5**, 29–35.

Ott, S. (1987). Sexual reproduction and developmental adaptations in *Xanthoria parietina*. *Nordic Journal of Botany*, **7**, 219–28.

Parsons, R. and Day, D. A. (1990). Mechanisms of soybean nodule adaptations to different oxygen pressure. *Plant Cell and Environment*, **13**, 501–12.

Pate, J. S. (1989). Synthesis, transport and utilisation of products of symbiotic nitrogen fixation. In *Plant nitrogen metabolism* (ed. J. E. Poulton, J. T. Romeo, and E. E. Conn), pp. 65–82. Plenum Press, New York.

Pearse, P. D., and Bauchop, T. (1985). Glycosidases of the rumen anaerobic fungus *Neocallimastix frontalis* grown on cellulosic substrates. *Applied and environmental Microbiology*, **49**, 1265–9.

Perasso, R., Baroni, A., Qu, L. H., Bachellerie, J. P., and Adoutte, A. (1989). Origin of the algae. *Nature*, **339**, 142–4.

Perotto, S., VandenBosch, K. A., Butcher, G. W., and Brewin, N. J. (1991). Molecular composition and development of the plant glycocalyx associated with the peribacteroid membrane of pea root nodules. *Development*, **112**, 763–73.

Pirozynski, K. A. and Malloch, D. W. (1975). The origin of land plants: a matter of mycotrophism. *BioSystems* **6**, 153–64.

Plazinski, J., Franche, C., Liu, C.-C., Lin, T., Shaw, W., Gunning, B. E. S., and Rolfe, B. G. (1988). Taxonomic status of *Anabaena azollae*: an overview. *Plant and Soil*, **108**, 185–190.

Pocock, K., and Duckett, J. G. (1985). On the occurrence of branched and swollen rhizoids in British hepatics: their relationship with the substratum and associations with fungi. *New Phytologist*, **99**, 281–304.

Poelt, J. (1970). Das Konzept der Artenpaare bei den Fechten. *Vortschrift Botanische Gesicht*, **4**, 187–98.

Provasoli, L., Yamasu, T., and Manton, I. (1968). Experiments on the resynthesis of symbiosis in *Convoluta roscoffensis* with different flagellate cultures. *Journal of the marine biological Association UK*, **48**, 465–79.

Pyke, K. A., and Leech, R. M. (1992). Chloroplast division and expansion is radically altered by nuclear mutants in *Arabidopsis thaliana*. *Plant Physiology*, **99**, 1005–8.

Rai, A. N., Rowell, P., and Stewart, W. D. P. (1981). ^{15}N incorporation and metabolism in the lichen *Peltigera aphthosa* Willd. *Planta*, **152**, 544–52.

Rands, M. L., Douglas, A. E., Loughman, B. C., and Hawes, C. R. (1992). The pH of the perisymbiont membrane in the green hydra–*Chlorella* symbiosis: an immunocytochemical study. *Protoplasma*, **170**, 90–3.

Read, D. J. (1983). The biology of mycorrhiza in the Ericales. *Canadian Journal of Botany*, **61**, 985–1004.

Read, D. J. (1984). The structure and function of the vegetative mycelium of mycorrhizal roots. In *The ecology and physiology of the fungal mycelium* (ed. D. H. Jennings and A. D. M. Rayner), pp. 215–40. Cambridge University Press.

Rees, T. A. V., and Ellard, F. M. (1989). Nitrogen conservation and the green hydra symbiosis. *Proceedings of the Royal Society of London B*, **236**, 203–12.

Robertson, J. G., Warburton, M. P., Lyttleton, P., Fordyce, A. M., and Bullwant, S. (1978). Membranes in lupin root nodules. II. Preparation and properties of peribacteroid membranes and the bacteroid envelope inner membranes from developing lupin nodules. *Journal of Cell Science*, **30**, 151–74.

Robinson, R. K. (1972). The production by roots of *Calluna vulgaris* of a factor inhibitory to growth of some mycorrhizal fungi. *Journal of Ecology*, **60**, 219–24.

Rosswall, T. (1983). The nitrogen cycle. In *The major biogeochemical cycles and their interactions, Scope 21* (ed. B. Bolin and R. B. Cook), pp. 46–50. John Wiley, Chichester.

Roth, L. E., Jeon, K., and Stacey, G. (1988). Homology in endosymbiotic systems: the term 'symbiosome'. In *Molecular genetics of plant–microbe Interactions* (ed. R. Palacios and D. P. S. Verma), pp. 220–5. APS Press, St Paul, Minnesota.

Rowan, R. and Powers, D. A. (1991). A molecular genetic classification of zooxanthellae and the evolution of animal–algal symbioses. *Science*, **251**, 1348–51.

Ruby, E. G. and McFall-Ngai, M. J. (1990). Morphological and physiological differentiation in the luminous bacterial symbionts of *Euprymna scolopes*. *Endocytobiology*, **4**, 323–5.

Schlaman, H. R. M., Okker, R. J. H., and Lughenberg, B. J. J. (1992). Regulation of nodulation gene expression by NodD in rhizobia. *Journal of Bacteriology*, **174**, 5177–82.

Scrivener, A. M., Slaytor, M., and Rose, H. A. (1989). Symbiont-independent digestion of cellulose and starch in *Panesthia cibrata* Saussure, an Australian woodeating cockroach. *Journal of Insect Physiology*, **35**, 935–41.

Sheehy, J. E. (1989). How much dinitrogen fixation is required in grazed glassland? *Annals of Botany*, **64**, 159–61.

Sleigh, M. A. (1989). *Protozoa and other protists*. Edward Arnold, London.

Smith, D. C. (1980). Mechanisms of nutrient movement between the lichen symbionts. In *Cellular interactions in symbiosis and parasitism* (ed. C. B. Cook, P. W. Pappas, and E. D. Rudolph), pp. 197–227. Ohio State University Press, Columbus.

Smith, D. C. and Douglas, A. E. (1987). *The biology of symbiosis*. Edward Arnold, London.

Smith, S. E., and Smith, F. A. (1990). Structure and function of the interfaces in biotrophic symbioses as they relate to nutrient transport. *New Phytologist*, **114**, 1–38.

Sprent, J. I. (1986). *The ecology of the nitrogen cycle*. Cambridge University Press.

Sprent, J. I., and Sprent, P. (1990). *Nitrogen-fixing organisms: Pure and applied aspects*. Chapman and Hall,

Starnes, S. M., Lambert, D. H., Maxwell, E. S., Stevens, S. E., Porter, R. D., and Stively, J. M. (1985). Cotranscription of the large and small genes of ribulose-1, 5-bisphosphate carboxylase/oxygenase in *Cyanophora paradoxa* . *FEMS Microbiology Letters*, **28**, 165–9.

Stevens, C. E. (1988). *Comparative physiology of the vertebrate digestive system*. Cambridge University Press.

Stewart, C. S. and Bryant, M. P. (1988). The rumen bacteria. In *The rumen microbial ecosystem* (ed. P. N. Hobson), pp. 21–74. Elsevier, London.

Streeter, J. G. (1989). Estimation of ammonium concentration in the cytosol of soybean nodules. *Plant Physiology*, **90**, 779–82.

Stubblefield, S. P., and Taylor, T. N. (1988). Recent advances in palaeomycology. *New Phytologist*, **108**, 3–25.

Stubblefield, S. P., Taylor, T. N., and Trappe, J. M. (1987). Vesicular–arbuscular mycorrhizae from the Triassic of Antarctica. *American Journal of Botany*, **74**, 1904–11.

Sutton, D. C., and Hoegh-Guldberg, O. (1990). Host–zooxanthella interactions in four temperate marine invertebrate symbioses: assessment of effect of host extracts on symbionts. *Biological Bulletin*, **175**, 178–86.

Thacker, E. J., and Brandt, C. S. (1955). Coprophagy in the rabbit. *Journal of Nutrition*, **55**, 375–85.

Thorsness, P. E., and Fox, T. D. (1990). Escape of DNA from mitochondria to the nucleus in *Saccharomyces cerevisiae*. *Nature*, **346**, 376–9.

Trappe, J. M. (1987). Phylogenetic and ecological aspects of mycotrophy in the angiosperms from an evolutionary standpoint. In *Ecophysiology of VA-mycorrhizal plants* (ed. C. R. Safir), p. 1–25. CRC Press, Boca Raton, Florida.

Trench, R. K. (1971). The physiology and biochemistry of zooxanthellae symbiotic with marine coelenterates. *Proceedings of the Royal Society of London B*, **177**, 225–64.

Truchet, G., Barkner, D. G., Camut, S., de Billy, F., Vasse, J., and Hugnet, T. (1989). Alfalfa nodulation in the absence of *Rhizobium*. *Molecular and general Genetics*, **219**, 65–8.

Udvardi, M. K., and Day, D. A. (1990). Ammonia (^{14}C methylamine) transport across the bacteroid and peribacteroid membranes of soybean root nodules. *Plant Physiology*, **94**, 71–76.

Udvardi, M. K., Price, G. C., Gresshoff, P. M., and Day, D. A. (1989). A dicarboxylate transporter on the peribacteroid membrane of soybean nodules. *FEBS Letters*, **231**, 36–40.

Van Bruggen, J. J. A., van Rens, G. L. M., Geertman, E. J. M., Zwart, K. B., Stumm, C. K., and Vogels, G. D. (1988). Isolation of a methanogenic endosymbiont of the sapropelic amoeba *Pelomyxa palustris* Greef. *Journal of Protozoology* **35**, 20–23.

Veira, D. M. (1986). The role of ciliate protozoa in the nutrition of the ruminant. *Journal of Animal Science*, **63**, 1547–60.

Veivers, P. C. Musca, N. Y., O'Brien, R. W., and Slayton, M. (1982). Digestive enzymes of the salivary glands and gut of *Mastotermes darwiniensis*. *Insect Biochemistry*, **12**, 35–40.

Veivers, P. J., Muhlemann, M., Slaytor, M., Leuthold, R. H., and Bignell, D. E. (1991) Digestion, diet and polyethism in two fungus-growing termites: *Macrotermes subhyalnis* Rambus and *Macrotermes michaelsoni* Sjostedt. *Journal of Insect Physiology*, **37**, 675–82.

Villareal, T. A. (1987). Evaluation of nitrogen fixation in the diatom genus *Rhizosolenia* Ehr. in the absence of its cyanobacterial symbiont *Richelia intracellularis* Schmidt. *Journal of Plankton Research*, **9**, 965–71.

Villareal, T. A. (1991). Nitrogen fixation by the cyanobacterial symbiont of the diatom genus *Hemiaulus*. *Marine Ecology Progress Series*, **76**, 201–4.

Vitousek, P. M., and Walker, L. R. (1989). Biological invasion by *Myrica faya* in Hawaii: plant demography, nitrogen fixation, ecosystem effects. *Ecological Monographs*, **59**, 247–65.

Vitousek, P. M., Matson, P. A., and van Cleve, K. (1989). Nitrogen availability and nitrification during succession: primary, secondary and old-field seres. *Plant Soil*, **115**, 229–39.

Wassman, C. C., Loffelhardt, W., and Bohnert, H. J. (1987). Cyanelles: organization and molecular biology. In *The Cyanobacteria—a comprehensive review* (ed. P. Fay and C. van Baleen), pp. 303–24. Elsevier, Amsterdam.

Waterbury, J. B., Calloway, C. B., and Turner, R. D. (1983). A cellulolytic nitrogen-fixing bacterium cultured from the gland of Deshayes in shipworms (Bivalvia: Teredinidae). *Science*, **221**, 1401–3.

Wellington, G. M. (1982). An experimental analysis of the effects of light and zooplankton on coral zonation. *Oecologia*, **52**, 311–20.

Whatley, J. M., and Whatley, F. R. (1981). Chloroplast evolution. *New Phytologist*, **87**, 233–47.

Whitehead, L. F., and Douglas, A. E. (1993). Populations of symbiotic bacteria in the parthenogenetic pea aphid (*Acyrthosiphon pisum*). *Proceedings of the Royal Society of London*, in press.

Whitehead, L. F., Wilkinson, T. L., and Douglas, A. E. (1992). Nitrogen recycling in the pea aphid (*Acyrthosiphon pisum*) symbiosis. *Proceedings of the Royal Society of London B*, **250**, 115–17.

Whittaker, R. J. (1975). *Communities and ecosystems*. Macmillan, London.

Whittaker, R. J., Bush, M. B., and Richards, W. (1989). Plant recolonisation and vegetation succession on Krakatau islands, Indonesia. *Ecological Monographs*, **59**, 59–123.

Woese, C. R. (1987). Bacterial evolution. *Microbiological Reviews*, **51**, 221–71.

Woese, C. R., Kandler, O., Wheelis, M. L. (1990). Towards a natural system of organisms: proposal for the domains Archaea, Bacteria and Eucarya. *Proceedings of the National Academy of Sciences USA*, **87**, 4576–9.

Wood, T., Bormann, F. H., and Voight, G. K. (1984). Phosphorus cycling in a northern hardwood forest, biological and chemical control. *Science*, **223**, 391–3.

Wren, H. N., Johnson, J. L., and Cochran, D. G. (1989). Evolutionary inferences from a comparison of cockroach nuclear DNA and DNA from their fat-body and egg endosymbionts. *Evolution*, **43**, 276–81.

Yamin, M. A., and Trager, W. (1979). Cellulolytic activity of an axenically culti-
vated termite flagellate, *Trichomitopsis termopsidis*. *Journal for general Microbiology*, **113**, 417–20.
Young, J. P. W., and Johnston, A. W. B. (1989). The evolution of specificity in the
legume–rhizobium symbiosis. *Trends in Ecology and Evolution*, **4**, 341–9.
Young, J. P. W., and Wexler, M. (1988). Sym plasmid and chromosomal genotypes
are correlated in field populations of *Rhizobium leguminosarum*. Journal for
general Microbiology, **134**, 2731–9.
Young, J. P. W., Downer, H. L., and Eardly, B. D. (1991). Phylogeny of the
phototrophic rhizobium strain BTAil by polymerase chain reaction-based
sequencing of a 16S rRNBA gene segment. *Journal of Bacteriology*, **173**, 2271–7.

Index